Smallholder Agriculture and Market Participation

Praise for this book

'Poole swims against the tide of meta-analysis, thematic review and non-differentiation. In this new work he embraces Sen's livelihoods entitlement framework anew and re-affirms people as central to market inclusion. For students of the successes and failures of agricultural markets as a tool to meet global development goals he provides cogent argument, up to date references and important guidance for practice.'

Ben Bennett, Deputy Director of the Natural Resources Institute and Professor of International Trade and Marketing Economics at the University of Greenwich

Smallholder Agriculture and Market Participation

Nigel Poole

Published by
Food and Agriculture Organization of the United Nations
and Practical Action Publishing 2017

Food and Agriculture Organization of the United Nations

Practical Action Publishing Ltd
The Schumacher Centre, Bourton on Dunsmore, Rugby, Warwickshire, CV23 9QZ, UK
www.practicalactionpublishing.org

© FAO, 2017

The authors have asserted their rights under the Copyright Designs and Patents Act 1988 to be identified as authors of this work. Such assertions shall not be construed as constituting a waiver of the privileges or immunities of FAO.

All rights reserved. No part of this publication may be commercially reprinted or reproduced or utilized in any form or by any electronic, mechanical, or other means, now known or hereafter invented, or in any information storage or retrieval system, without the written permission of the publishers.

Product or corporate names may be trademarks or registered trademarks, and are used only for identification and explanation without intent to infringe.

A catalogue record for this book is available from the British Library.
A catalogue record for this book has been requested from the Library of Congress.

ISBN 978-1-85339-941-1 Paperback
ISBN 978-1-85339-940-4 Hardback
ISBN 978-1-78044-941-8 ebook
ISBN 978-1-78044-940-1 Library pdf

Citation: Poole, N. (2017) *Smallholder Agriculture and Market Participation*, Rugby, UK: Practical Action Publishing, <http://dx.doi.org/10.3362/9781780449401>

Since 1974, Practical Action Publishing has published and disseminated books and information in support of international development work throughout the world. Practical Action Publishing is a trading name of Practical Action Publishing Ltd (Company Reg. No. 1159018), the wholly owned publishing company of Practical Action. Practical Action Publishing trades only in support of its parent charity objectives and any profits are covenanted back to Practical Action (Charity Reg. No. 247257, Group VAT Registration No. 880 9924 76).

FAO encourages the use, reproduction and dissemination of material in this information product. Except where otherwise indicated, material may be copied, downloaded and printed for private study, research and teaching purposes, or for use in non-commercial products or services, provided that appropriate acknowledgement of FAO as the source and copyright holder is given and that FAO's endorsement of users' views, products or services is not implied in any way.

All requests for translation and adaptation rights, and for resale and other commercial use rights should be addressed to www.fao.org/contact-us/licence-request or to copyright@fao.org.

FAO information products are available on the FAO website (www.fao.org/publications) and can be purchased through publications-sales@fao.org

FAO ISBN 978-92-5-109939-1

The designations employed and the presentation of material in this information product do not imply the expression of any opinion whatsoever on the part of the Food and Agriculture Organization of the United Nations (FAO) concerning the legal or development status of any country, territory, city or area or of its authorities, or concerning the delimitation of its frontiers or boundaries. The mention of specific companies or products of manufacturers, whether or not these have been patented, does not imply that these have been endorsed or recommended by FAO in preference to others of a similar nature that are not mentioned. The views expressed in this information product are those of the author(s) and do not necessarily reflect the views or policies of FAO, or of Practical Action Publishing Ltd or its parent charity Practical Action. Reasonable efforts have been made to publish reliable data and information, but the authors and publisher cannot assume responsibility for the validity of all materials or for the consequences of their use.

Cover design: Andrew Corbett
Cover photo shows small-scale cassava traders awaiting business in an urban market, northern Zambia. Credit: the author.
Typeset by vPrompt eServices, India
Printed in the United Kingdom by Short Run Press Ltd

Contents

About the author ix

Part I
The potential and challenges of smallholder agriculture 1

1. Introducing smallholder agriculture 3
   The significance of smallholder agriculture 3
   The contribution of agricultural development
   to reducing hunger and poverty 8
   Challenges in fostering agricultural entrepreneurship 12
   Commercializing agriculture 16
   The limitations of agricultural development for poverty reduction ... 25
   A note on terminology 27
   About the book 29

2. Policy approaches and theoretical considerations 35
   Changing policies: from market systems to value chains 35
   Theoretical approaches to households, markets,
   and marketing for the poor 42
   Transaction costs 45
   Households and livelihoods 48
   Value chain thinking 53

3. Financial services for agricultural smallholders 65
   Background 65
   Finance and transaction costs 68
   Financial innovations for farming 73

4. Risk management for agricultural smallholders 81
   Introduction 81
   Assessing risk in rural value chains 83
   Risk management 87
   Agricultural insurance for agricultural development 97
   Economic risk and types of value chains 101
   Conclusions 110

http://dx.doi.org/10.3362/9781780449401.000

Part II
Case studies of smallholder agriculture 115

5. Introduction to Part II: Assessing the impact of commodity development projects on smallholder participation in agricultural markets: Case studies from Ethiopia, Peru, Tanzania, and Zambia 117
   Introduction 117
   Cases 118
   Research approach and framework 121
   Methodology 124
   Summary of lessons learned 126

6. A diversification programme for vegetable exports in Ethiopia 129
   Project context 129
   Findings 131
   Conclusions 136

7. Production and commercialization of oilseeds in Peru 139
   Project context 139
   Findings 140
   Conclusions 145

8. Sisal product and market development in Tanzania 149
   Project context 149
   Findings 150
   Conclusions 154

9. Strengthening the productivity and competitiveness of smallholder dairying in Zambia 157
   Project design and implementation 157
   Project context 157
   Findings 159
   Conclusions 162

10. Assessing the impact of projects on smallholder participation in agricultural markets: Synthesis and conclusions 165
    Recapitulating the theory of change: expected relationships 165
    Lessons learned 166
    Conclusions 178

11. Postscript: 'Going local' with development policies	183
Feeding the world	183
A macro perspective	185
A sectoral perspective	186
Policy interventions and tensions of scale: going micro	188
Index	193

About the author

Nigel Poole originally trained in natural sciences at the University of Nottingham and in agricultural extension at the University of Reading, UK. He began working in overseas agriculture in Swaziland and then moved to Paraguay in South America for 11 years, where some of his best days were spent being a cowboy. He returned to the UK and switched into socioeconomics, studying for Masters and PhD degrees in agricultural economics at Wye College, University of London.

Since 2007 he has worked in the Centre for Development, Environment and Policy at SOAS, University of London, where he is Professor of International Development. He has research experience in diverse countries in Latin America, the Mediterranean Basin, Sub-Saharan Africa and South and South-east Asia. His teaching is mainly in marketing, business and knowledge management.

He is currently researching agrifood and nutrition value chain linkages with the DFID-funded multi-institutional research programme 'Leveraging Agriculture for Nutrition in South Asia' (LANSA) covering Bangladesh, India, Pakistan and Afghanistan (http://www.lansasouthasia.org/). He is leading the Afghanistan Working Group and the agrifood-nutrition value chain studies in the overall programme.

He is also Chairman of the Board of Directors, CATIE (Centro Agronómico Tropical de Investigación y Enseñanza/Centre for Tropical Agricultural Research and Teaching), Costa Rica. In his spare time, among other things, he is Chairman of a small charity, the Wye and Brook India Trust (registered charity number 288217), supporting schooling in Delhi, India.

PART I
The potential and challenges of smallholder agriculture

CHAPTER 1
Introducing smallholder agriculture

This introductory chapter discusses the importance of smallholder agriculture for economic development, poverty reduction, and livelihood enhancement in developing countries.

We note that agriculture is an occupation as well as an economic enterprise, and that women are much involved throughout the developing world. Agriculture is intimately linked with food security, health, and nutrition through direct consumption and market linkages. Agriculture also affects other dimensions of development such as environmental management and sustainability.

The participation of farmers in markets is an important determinant of well-being and development. The chapter continues by analysing the opportunities for, and weaknesses of, smallholders engaging in commercial agriculture, and how the particular and local circumstances of such farmers mean that they cannot be treated as a homogeneous group. It suggests that there are limits to policies of social and economic inclusion as a means to overcome inequality, and notes that agricultural development cannot solve all complex development problems. Targeting development initiatives is therefore a challenge which is explored later in the book.

The final section clarifies a variety of terms that are used to discuss development issues and policy, and then explains the origin, context, and purpose of the book.

Keywords: smallholder farmers, agricultural development, poverty reduction, market participation, policies, Sustainable Development Goals

The significance of smallholder agriculture

Agriculture is a major factor in sustainable global resource management in terms of land, water, biodiversity, and emissions to land, water, and the atmosphere. Agriculture is not just rural, but occurs on and within the margins of urban areas and major cities. Production depends on resource-intensive input supplies, and interfaces with other economic sectors such as transport, energy, water, labour, and commodity markets through processing, manufacturing, and distribution of products. The politics of food is not just about fair prices for producers and consumers. Agriculture is about international development and poverty reduction, and plays a fundamental part in environmental management and in the human causes of, and responses to, climate change. A glance at the Sustainable Development Goals, adopted by the international community in 2015 and set as targets for 2030, will confirm that agriculture is linked to many sectors within the development agenda.

http://dx.doi.org/10.3362/9781780449401.001

Smallholder agriculture is one of the principal economic occupations in the world and is the main source of income and employment for the 70 per cent of the world's poor who live in rural areas. Smallholder households account for 60 per cent of global agriculture. Many small-scale producers are women with multiple responsibilities besides farming. Most farming households produce a diverse range of farming products – different crops and livestock which fit into the home economy in different ways. They are involved in other economic activities besides farming so that, despite its significance, agriculture is just one of a number of diverse and competing sources of livelihood support.

The latest World Bank report on enhancing global agribusiness opens with the following statement:

> Sustainable agricultural development is one of the most powerful tools to end extreme poverty and boost shared prosperity. Agriculture is the economic and social mainstay of some 500 million smallholder farmers, and in developing countries, the sector is the largest source of incomes, jobs and food security. Sustainable, inclusive growth in the agriculture and food sectors creates jobs – on farms, in markets, cities, towns and villages, and throughout the farm-to-table food production and consumption chain.
>
> Seen against the backdrop of an increasing world population that is expected to reach nine billion by 2050, rising food demand is estimated to increase by at least 20% globally over the next 15 years with the largest increases projected in sub-Saharan Africa, South Asia and East Asia. Boosting the productivity, profitability and sustainability of agriculture is essential for fighting hunger and poverty, tackling malnutrition and boosting food security. In short, the world needs a food system that can feed every person, every day, everywhere with a nutritious and affordable diet, delivered in a climate-smart, sustainable way.
>
> To achieve this goal, we need to be more productive and efficient in the way we grow food, while building the resilience of both farmers and food supply chains while simultaneously reducing the environmental footprint of the agriculture and food sectors. This process requires policies and regulations that foster growth in the agriculture and food sectors, well-functioning markets, and thriving agribusinesses that make more food available in rural and urban spaces (World Bank, 2017: v).

More than economics

Smallholder agriculture is more than an economic activity. For some, perhaps for many, it is a way of life:

> Agriculture in the Pacific is more than the occupation of the great majority of people; it is their satisfaction, the means by which what

> **Box 1.1 Smallholders and family farmers**
>
> **Environment**: Eighty per cent of the farmland in sub-Saharan Africa and Asia is managed by smallholders (working on up to 10 hectares). While 75 per cent of the world's food is generated from only 12 plant and five animal species, making the global food system highly vulnerable to shocks, biodiversity is key to smallholders whose systems keep many rustic and climate-resilient varieties and breeds alive.
>
> **Economy**: Out of the 2.5 billion people in poor countries living directly from the food and agriculture sector, 1.5 billion people live in smallholder households. Many of those households are extremely poor: overall, the highest incidence of workers living with their families below the poverty line is associated with employment in agriculture.
>
> **Social**: Women comprise an average of 43 per cent of the agricultural labour force of developing countries – up to almost 50 per cent in East and Southeast Asia and sub-Saharan Africa. Should women farmers have the same access to productive resources as men, they could increase yields on their farms by 20–30 per cent, lifting 100–150 million people out of hunger. Women are the quiet drivers of change towards more sustainable production systems and a more varied and healthier diet.
>
> **Governance**: Smallholders provide up to 80 per cent of the food supply in Asia and sub-Saharan Africa. Their economic viability and contribution to diversified landscape and culture are threatened by competitive pressure from globalization and integration into common economic areas; their fate is either to disappear and become purely self-subsistence producers, or to grow into larger units that can compete with large, industrialized farms.
>
> *Source*: summarized from FAO (2012).

survives of tradition is largely expressed and maintained, and the channel of individual creativity and enterprise within traditionally close confines of the extended family and community (Barry Weightman, 1989, quoted in Rogers et al., 2010: 2).

Box 1.1 explores some dimensions of the type of farming we are considering.

Smallholder agriculture is most important of all for its contribution to the food security, nutrition, and health of many poor people:

> Small-scale agriculture is the main source of food in the developing world, producing up to 80 percent of the food consumed in many developing countries, notably in sub-Saharan Africa and Asia. With poor rural households making up two-thirds of the global population earning less than $1.25 per day, smallholder agriculture is also an important source of income underpinning the livelihoods of vast numbers of poor people. Smallholders and small family farms are therefore central to an inclusive development process and their contribution is crucial to food security (Arias et al., 2013: 6).

The significance of agriculture and the rural economy to poverty reduction can be illustrated with statistics from the World Bank's World Development Indicators. In developing countries and regions, the rural population is relatively high, and agriculture accounts for a relatively large but declining

Table 1.1 Rural indicators – major regions of the world

	Rural population			Agriculture	
	% of total growth		Annual % growth	% of GDP	
	2000	2014	2014	2000	2014
World	53	47	0.2	5	4
East Asia and Pacific	59	44	−1.4	8	5
Europe and Central Asia	32	29	−0.2	3	2
Latin America and Caribbean	25	20	−0.3	6	5
Middle East and North Africa	41	36	1	9	6
North America	21	19	−0.1	1	1
South Asia	73	67	0.7	23	18
Sub-Saharan Africa	69	63	1.9	20	17
Low income	75	70	2.2	34	31
Lower middle income	67	61	0.7	22	16
Upper middle income	50	37	−1.5	10	7
High income	23	19	−0.7	2	1

Source: World Bank (2016)

proportion of gross domestic product (GDP). Food production has increased significantly in some regions in the last decade, but not equally in all, and further increases are necessary considering likely population growth rates, particularly in sub-Saharan Africa. Agricultural productivity in developing regions has risen during the period 2000–2014, but is still much lower than in more developed regions (Table 1.1 and Table 1.2).

At the same time, smallholder agriculture has many defining characteristics and faces many obstacles in reaching its potential contribution to development and poverty reduction. This volume focuses on mechanisms for enhancing economic access to markets:

> Smallholder agriculture is characterized by small production volumes of variable quality that reflect limited access to inputs and finance, low levels of investment and limited access to, and knowledge of, improved agricultural technologies and practices. High levels of price and production risk and uncertainty and limited access to tools to manage them deter investment in more productive new technologies that would enable smallholders to produce surpluses for sale in markets. Inadequate infrastructure, high costs of storage and transportation and non-competitive markets also militate against production of a marketable surplus (Arias et al., 2013: 6).

While the economic output of many countries in developing regions is highly agricultural, the health and nutrition indicators for some of these countries are among the poorest in the world; sub-Saharan Africa and South Asia are the regions with the most significant deprivation. Were the data in

Table 1.2 Food production and productivity indicators – major regions of the world

	Food production index		Agricultural productivity	
	2004–2006 = 100		Agricultural value added per worker	
	2000	*2013*	*2000*	*2014*
World	88.6	123	1,562	2,161
East Asia and Pacific	84.8	127.7	1,095	1,738
Europe and Central Asia	96.5	110.5	8,709	14,026
Latin America and Caribbean	83.0	128.5	4,627	7,007
Middle East and North Africa	81.6	117.3	4,067	6,194
North America	95.5	113.5	44,132	78,230
South Asia	90.9	132.3	848	1,087
Sub-Saharan Africa	82.4	129.6	771	1,207
Low income	83.2	133.4	396	510
Lower middle income	85.6	130.7	1,075	1,577
Upper middle income	84.0	84.0	128	1,314
High income	98.5	107.7	22,075	38,321

Source: World Bank (2016)

Table 1.3 to be disaggregated by age and gender, the under-fives and adolescent girls would register significantly.

Digging a little deeper into the data in Table 1.3, it is evident that the Oceania region – including many Small Island Developing States (SIDS) – registers high levels of stunting among children, comparable to the levels in Africa. It is more surprising to observe the high rates of adult obesity in the Caribbean and Oceania, which encompass most of the SIDS. These are comparable to the levels of obesity in many wealthy countries in developed regions of the world. These data illustrate the phenomenon of the double-burden of malnutrition, which is under- and over-nutrition. The co-existence of stunting, micronutrient deficiencies, and adult obesity creates major challenges for local and global food systems. The linkages between agriculture and nutrition are currently high on the international research agenda: for example, in the UK Government-funded research programme 'Leveraging Agriculture for Nutrition in South Asia' (LANSA, undated). Recent publications taking a value perspective on the linkages are Devaux et al. (2016) and Maestre et al. (2017).

Because of its contribution to food supplies and nutrition, smallholder agriculture is also an important factor supporting social and political stability. For example, high onion prices in India in 1998 led to the fall of the BJP government, and continue to act as a bellwether of political tension. Rising food prices are implicated, among other factors, in fomenting the Arab Spring (Breisinger et al., 2012). In all countries, food price rises make a significant contribution to the rate of inflation, to which poor people are most sensitive.

Table 1.3 Nutrition-related health indicators – major regions of the world

	Prevalence of stunting among children (%)	Prevalence of micronutrient deficiencies and anaemia among children (%)			Prevalence of obesity among adults (%)
		Anaemia	Vitamin A deficiency	Iodine deficiency	
	Most recent observation	Most recent observation			2008
World	25.7	47.9	30.7	30.3	11.7
Countries in developed regions	7.2	11.8	3.9	37.7	22.2
Countries in developing regions	28.0	52.4	34.0	29.6	8.7
Africa	35.6	64.6	41.9	38.2	11.3
Southern Asia	45.5	66.5	50.0	36.6	3.2
Caribbean	6.7	41.3	17.8	59.8	20.3
Oceania ex Australia/New Zealand	35.5	53.8	11.6	31.8	22.4

Source: FAO (2013: Annex 1)

The contribution of agricultural development to reducing hunger and poverty

Development goals

Since 2000, the international commitment to agricultural development has been tackled within the framework of the Millennium Development Goals (MDGs), the high level objectives with targets and indicators which ran up to 2015. Agricultural development is a strategy which is usually considered to have addressed MDG 1, that is, the eradication of extreme poverty and hunger. Through increasing rural incomes, agricultural development has contributed, between the years 1990 and 2015, towards halving the proportion of people whose income is less than one dollar a day.

By increasing opportunities, agricultural development can also help to achieve full and productive employment and decent work for all, including women and young people; and through boosting income, employment, and food production and lowering prices, agricultural development can help to halve the proportion of people who suffer from hunger. These multiplier effects of agricultural development are well understood. Indeed, agricultural development has conceivably helped to address, inter alia, reductions in child mortality, improving maternal health, and combatting HIV/AIDS, malaria, and other diseases.

The framework of the Sustainable Development Goals (SDGs) makes reference to more precise goals and indicators relating to equity, gender, and sustainability issues (see UN Sustainable Development Knowledge Platform; UN-DESA, undated). We can still expect that agricultural development – critical to many of the world's poorest people – will have a major contribution to the new SDG era.

Poverty is not just rural, nor is development just economic, nor do international goals encapsulate all the desirable objectives for global society. Economic incentives, new knowledge about the world, and new technical skills, to name just three external influences, act upon individuals, households, and communities to change culture and values in various ways. It is obvious that other approaches and mechanisms, as well as appropriate agricultural development policies, are required to meet the SDGs: agricultural development may be necessary for achieving each objective, but it is certainly not sufficient.

The new SDGs are integrated and synergistic – and perhaps antagonistic: handling the trade-offs between agriculture, economic development, and environmental management, for example, will be challenging (Waage and Yap, 2015). Exploiting potential synergies is a positive approach to addressing the goals. Given its importance to the rural economy, where many poor people are located, agriculture can be expected to contribute directly to SDGs 1 and 2:

- Goal 1. End poverty in all its forms everywhere.
- Goal 2. End hunger, achieve food security and improved nutrition, and promote sustainable agriculture.

Interrelationships between goals

There are potential conflicts and trade-offs between goals, but smallholder agriculture can also address issues of equity, inclusion, gender and empowerment, health, employment, natural resources management, climate change, and sustainability. Figure 1.1 depicts all 17 SDGs, highlighting in the centre those on which agriculture has an immediate bearing. In the second ring are those on which agriculture can plausibly have a less direct but important impact.

Many development challenges are multifaceted, 'wicked' problems that are difficult to resolve because of inadequate information and knowledge, multiple and contradictory opinions, perceived high costs, and complexity. Recent research among policy stakeholders in Afghanistan shows how agricultural and other sector policies can be poorly designed and integrated (Poole et al., 2016b):

> Major barriers remain to leveraging agriculture for nutrition. These are not confined to the specific sectors but are symptomatic of Afghanistan's broader development challenges. Extreme dependence on external human and financial resources shapes policy and practice according to international expectations, but fails to deliver efficient and effective

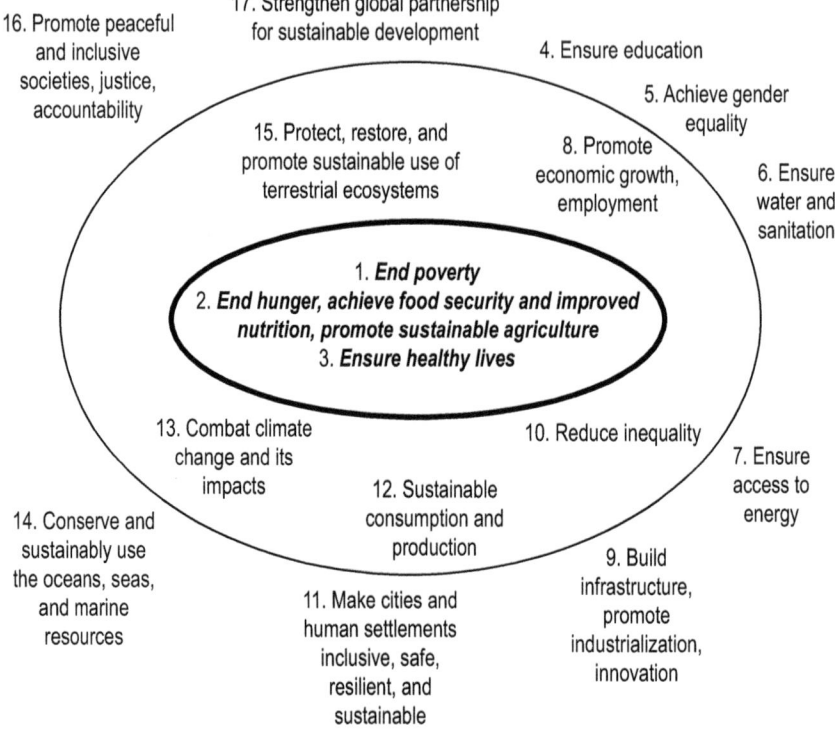

Figure 1.1 Agriculture and the Sustainable Development Goals
Source: adapted from Waage and Yap (2015)

processes and outcomes. In particular, we find that lack of capacity and resources within government Ministries and Departments along with poor infrastructure and huge security concerns remain major barriers to progress (Poole et al., 2016b: 88).

The complexity of wicked inter- or multi-sectoral problems can be illustrated by examining Figure 1.2, which is a representation, by no means comprehensive, of the interrelationships between primary agricultural production and other sectors. It captures both the upstream functions of the agricultural supply industry and the ecological functions of the natural environment, and a range of factors which affect the downstream value chain functions, right down to the consumption and health impact of agri-food products (Maestre et al., 2017). Here the focus is on food and health, and the factors are located approximately according to their position in the chain. Non-food agricultural products would constitute a branch into other industries, such as bioenergy – but are too difficult to depict in the same diagram.

But this does not mean that nothing can be done. We know that inherent in development are processes whereby natural, human, and economic

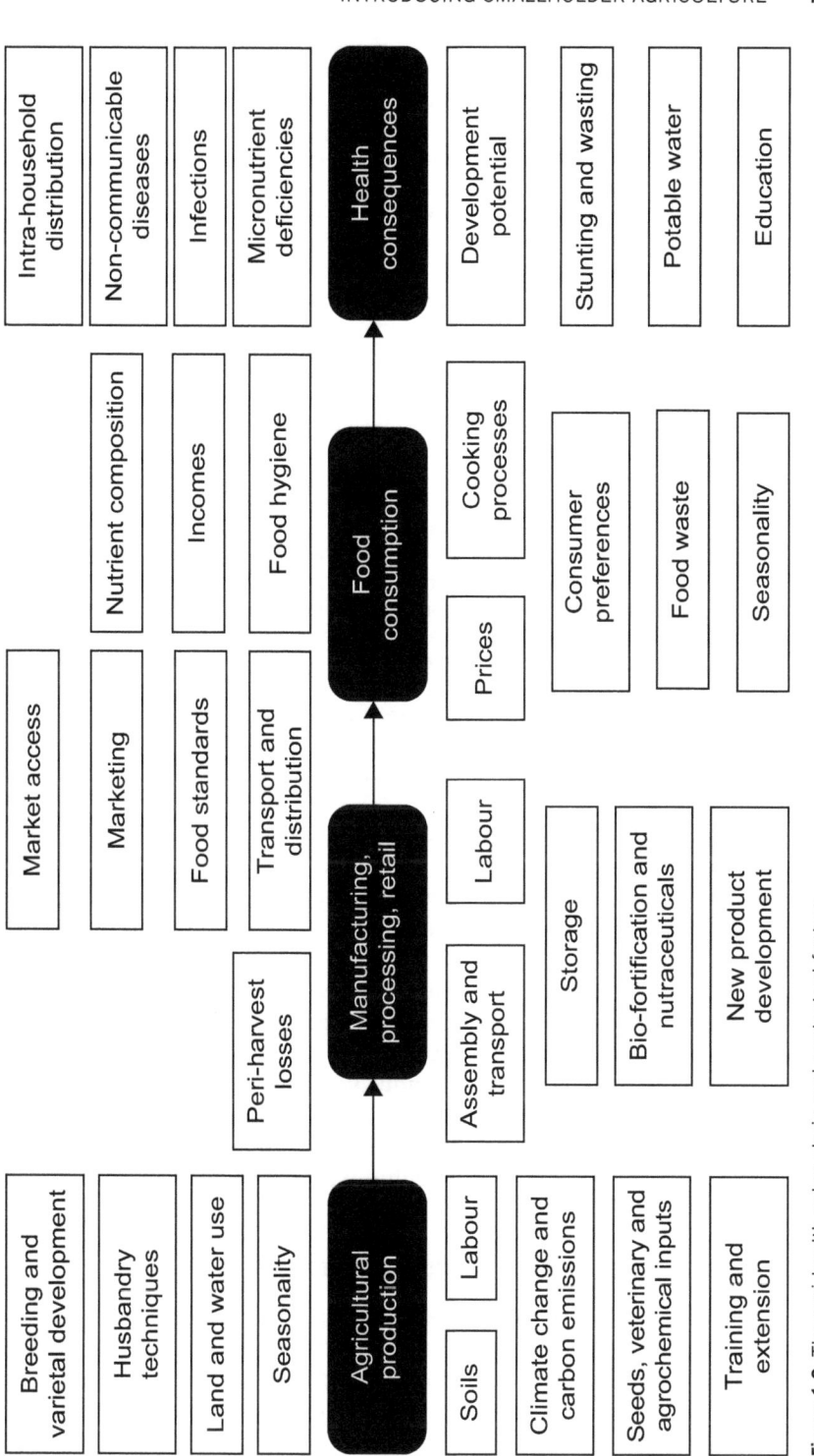

Figure 1.2 The agri-health value chain and contextual factors
Source: adapted from Waage et al. (2011)

> **Box 1.2 Livelihoods and ecology in the Sahel region of West Africa**
>
> A survey approach was used to examine the contribution to rural livelihoods in Burkina Faso of major tree products, derived from baobab (*Adansonia digitata*), shea (*Vitellaria paradoxa*) and *néré* (*Parkia biglobosa*). While home-produced cereals constitute the major part of the people's diets, tree products make a significant contribution to food security during the dry *soudure* season when cereal granaries are empty and the new harvest is pending. Tree products also served as a mechanism for managing household strategies for exchange of goods, marketing of products, and managing food consumption through the seasons. Understanding of the agroecological importance of trees and tree management was reinforced.
>
> Households were not all the same: rural heterogeneity down to intra-household level is an important feature of tree product utilization, with variations in tree management and product utilization depending on human and economic relationships as well as natural resource endowments.
>
> The complexity of and variation in human relationships was evident at multiple levels of the Burkinabé society: at the community level, where local authorities exercise a degree of control over natural resources; within the typical compound of households; among men and women and children within and between the constituent households.
>
> One of the principal implications of this research for enhancing the contribution of trees to sustainable livelihoods in the Sahel is that conservation and enhancement initiatives must be tailored to the local context. Implementation must be negotiated according to the important agroecological, human, and tree specificities, land tenure and fragmentation, particularly in the more fragile northern part of the country, and, at the same time, address the gender and age roles of different householders and authority figures, thus exploiting the potential for sustainable enterprise and greater productivity.
>
> *Source*: adapted from Poole et al. (2016a)

resources grow and are reallocated among livelihood activities within people's households. Recent research in West Africa has emphasized how the people of the Sahel and their skills are a resource that can be used to manage the competing claims of sustainable management of a fragile environment and secure rural livelihoods (Box 1.2).

The changes accompanying these processes involve shifts between productive sectors and spatial contexts, driven by the initiatives of entrepreneurs, augmented by external support, and facilitated by incentives from an external environment which is conducive to productive investment and returns. Within this broad context, both theory and practice are persuasive about the role of increasing rural peoples' access to remunerative markets for their products as a mechanism of agricultural development.

Challenges in fostering agricultural entrepreneurship

We have seen that agricultural development and entrepreneurship are key processes for improving the livelihoods of poor people, for wider economic development, and also make sense for ensuring food security. The thesis – or theory of change – is that the commercialization of smallholder agriculture is likely to stimulate primary production and thereby increase household and food security (at a local and regional scale), for increasing incomes of farmers and employees, for enabling and stimulating the supply of agricultural services and goods to rural areas, and for the secondary (agro-industrial) and tertiary (services) economic sectors. Moreover, social and environmental objectives can be met by promoting smallholder agriculture as part of a global food system that does not depend entirely on intensive agriculture.

Targeting development support

The fact that the global population of smallholder farmers is characterized by heterogeneity will lead to further differentiation and specialization in rural communities. Thus, not all smallholders will be able or willing to participate or benefit directly from the opportunities and processes of agricultural development. Targeting support will create opportunities and foster the inclusion of resource-poor people, but unequal starting points, such as the level of asset endowments for agricultural entrepreneurship, give rise to the evolving differences. The vulnerability and inequality of the most resource-poor people have to be handled by transfers and safety nets, plus other targeted opportunity-enhancing approaches. For example, appropriate education opens a pathway into higher-value employment in an inclusive local economy.

Both push and pull factors are at work in altering the rural population structure in developing economies. Migration from rural to urban areas and overseas is accelerating processes of feminization and ageing in the agriculturally active population that is also impacted by the prevalence of HIV/AIDS and other persistent and destructive phenomena. Remittances have become exceedingly important and constitute a major and irreplaceable flow of resources approaching $500 bn per annum (Sutherland, 2013), with around 250 million migrants financially supporting 1 billion people in their countries of origin. Nevertheless, there is an 'employment myth' which means that expectations of migrants are often frustrated by lack of opportunities in destination markets. At the same time, alternative local economic development and food production can be hit as human capital becomes stretched and climate change requires innovative responses to sustain and increase agricultural productivity.

Changing contexts

Two recent international reports highlight the immediate and longer-term challenges. *Rural Development Report 2016: Fostering Inclusive Rural Transformation* by the United Nations International Fund for Agricultural Development (IFAD) draws attention to the rapidly changing global context: the global economy, rates of urbanization and increasing food demand, climate change, erratic energy supplies, conflict, and increasing inequality (IFAD, 2016).

> **Box 1.3 Employment potential of agriculture**
>
> 'Employment creation in the formal sector falls far below the level required to absorb new market entrants. Moreover, most new entrants lack the skills they need to enter firms operating at higher levels of productivity and wages. As a result, the overwhelming majority of young people are destined for employment on farms, rural enterprises or in the informal sector ...
>
> 'Successful adaptation by smallholder farmers could dramatically reduce the risks posed by climate change. Innovations in water management and irrigation, drought-resistant seed strains, soil conservation, and new tillage and climate-resilient cropping patterns could all make a difference. Africa's farmers ... need help from their governments and the international community in scaling up adaptation ...
>
> 'Unlocking the productive potential of Africa's farmers would strengthen economic recovery. It would raise incomes, create jobs, create new markets, open new opportunities for investment, and link the farm and the rural non-farm economy with other growth centres.'
>
> *Source*: Africa Progress Panel (2012)

Likewise, the report of the United Nations Food and Agricultural Organization (FAO) *State of Food and Agriculture 2016: Climate Change, Agriculture and Food Security* highlights the urgency of responding to the challenges affecting agriculture and food production:

> Unless action is taken now to make agriculture more sustainable, productive and resilient, climate change impacts will seriously compromise food production in countries and regions that are already highly food-insecure. These impacts will jeopardize progress towards the key Sustainable Development Goals of ending hunger and poverty by 2030; beyond 2030, their increasingly negative impacts on agriculture will be widespread (FAO, 2016: v).

Strategies for prioritizing food security and environmental conservation may include maintaining the dwindling rural labour force and stimulating agriculture by providing production subsidies and investments in rural infrastructure. While many rural people want to move to the cities, it may be not possible for them all to do so. Prioritizing investment in the rural economy, therefore, is part of the solution to rural, urban, food, and employment problems (Box 1.3).

A people perspective

A particular challenge in many contexts is to engage a new generation of people in agriculture. In Samoa, for example, farming is said to be among the least valuable of the employment or career opportunities for a young villager, coming behind migration, trade, and fishing (Angelucci et al., undated). The age profile of smallholder farmers is steadily rising in many countries and regions because young people are leaving rural areas to seek opportunities elsewhere. Farming can become an attractive profession if smallholder farms can be transformed into modern businesses. For this, at least three things are needed (IFAD, 2011: 5):

- investment in social and economic infrastructure;
- remunerative economic opportunities for young people; and
- opportunities and appropriate skills development.

While migration does present attractive new opportunities to people with qualifications, integration into a vibrant rural economy can offer higher returns through agriculture and rural business. The demand for added-value agri-food products sourced from dynamic, innovative, and modern businesses within a profitable agricultural base will persist, and appropriate investments will be necessary. Human resources are critical: educational policies must be tailored to local needs with the intention of 're-skilling' a younger generation to overcome the barriers to local development and enabling the exploitation of new opportunities (Poole et al., 2013a).

Karembu (2013) goes further, arguing that high-technology agriculture and sophisticated skills are necessary for the development of Africa. She is concerned about increasing food demand: where the food will come from, how it will be produced, and who will produce it. The required productivity increases will depend on modern, scientifically based agricultural technologies. Yet outmigration of young people has left rural food production as the responsibility of an older generation who lack the technical and other capitals to innovate, adopt, and adapt to modern farming and market systems. She ponders likewise how young people can be attracted into farming given the prevailing view of agriculture as a painful and low-end labour market. Can farming become 'pleasurable and profitable with supportive infrastructure to make it exciting, worthwhile and recognised as an important contribution to modern society'?

In conclusion, she argues:

> Agriculturalists agree that the long-term sustainability of existing food production systems will largely depend on appropriate uptake and application of modern science and technologies. Education, empowerment and motivation of young people to take up agricultural activities are a prerequisite for improved and sustainable food production in Africa given their big numbers ... With better opportunities for access to technologies, entrepreneurial skills and social marketing, young people could funnel their youthful idealism, energy and determination into a positive force for change within the agricultural sector. This would ultimately result in sustainable production of the food required to support the growing population in Africa (Karembu, 2013: 97).

Commercializing agriculture

Understanding smallholder market participation

Pure subsistence is very rare, and cash needs for household expenditures are an almost universal reality. Diversification in patterns of economic production and capital accumulation are fundamental livelihood strategies for economic well-being. Thus, almost all people, including smallholder farmers and the poorest, are linked to markets in the wider non-agricultural economy. In the first instance, these are markets for consumer goods, including food. In rural areas, many farmers will source agricultural inputs such as seeds, implements, and fertilizers from local suppliers, or rent other agricultural services. Many people in rural contexts are also linked to labour markets. At the very least, some members of farming households often work as farm labourers or in secondary enterprises, or migrate out of rural areas to urban areas or further afield.

A decade ago, the World Bank's Development Report on agriculture set the agenda for commercialization and explored opportunities and weaknesses in the smallholder context (World Bank, 2007). National governments, international donors, and non-governmental organizations have adopted the commercialization narrative. Typical of public statements is the following:

> The official policy statements of many national governments in the Pacific accord a central role to the intensification and commercialization of smallholder agriculture as a means of stimulating the rural economy and alleviating poverty. Axiomatic to this stance is the belief that smallholder agriculture is uniquely positioned to deliver broad-based growth in rural areas (Rogers et al., 2010: 6).

Smallholder farmers contribute to agricultural exports of many high-value fruit and vegetable products, although large-scale agriculture dominates these markets. Smallholders contribute especially to the beverages sector: smallholder tea, coffee, and cocoa farmers are often linked closely to local agribusinesses and international trade. Incomes from such sources are critical for current household expenditures, while larger sums such as those from overseas remittances can contribute to capital investment. However, for staples and many other commodities, outside local markets, it is usually the better-off smallholders who are net sellers of agricultural produce.

Most smallholder sellers of agricultural products are in fact net buyers, because productive resources are limited and productivity is low. There is no doubt that smallholder agricultural productivity can increase, and has to increase, to enhance farmers' livelihoods and to meet increasing demands for food security. But the effectiveness of incentives and distribution of benefits from investment in new research and technology will be much attenuated if the market linkages are not improved at the same time: 'Raising smallholder productivity is obviously a strategic necessity, but attempts to raise productivity will have limited success if smallholder linkages to markets are not strengthened simultaneously. Similarly, strengthening market linkages will have little benefit with existing low levels of productivity' (Arias et al., 2013: 6).

According to Wiggins and Keats (2013), the contribution of smallholder agriculture to reducing poverty and hunger in low-income countries depends on sustainable access to markets. The primary focus of terms such as market participation and economic inclusion is to consider the multiple ways and the extent to which smallholder farmers are able to sell their output to buyers who may be itinerant assemblers and traders, agents for larger-scale procurement systems, traders in markets or direct consumers. Other markets besides product markets must also be considered as integral elements in local production systems:

> Increased market participation implies the transition from subsistence farming to a market engagement mode, whereby frequent use of markets is made for the purpose of exchanging products and services. Markets refer to both input markets for the exchange of factors of production and output markets for the exchange of agricultural products (Amrouk et al., 2013: 6).

Box 1.4 explores what market participation can mean.

Box 1.4 What do we mean by market participation?

Market participation is the ability of an entity to participate in a market efficiently and effectively. For our purposes it implies the transition by farmers from subsistence farming to a market engagement mode, whereby inputs are increasingly purchased and outputs sold off the farm to traders. It is a process as well as an outcome. The transition from subsistence, or from a lower to a higher level of market participation, is influenced by the ability of farmers to produce products which meet market expectations in terms of quality, standards, supply consistency, and ability to deliver products on time for sale at a viable price.

Most smallholder farmer market participation in developing countries is limited owing to factors which are both internal to the farmer or household, and external, from the surrounding environment. The internal factors are barriers which relate to the failure by farmers to meet market expectations due to lack of physical and financial assets, such as land and credit, and human assets such as skills, commercial contacts and labour, and even time. Smallholder farmers also frequently lack commercial information; physical infrastructure is poor causing high transaction costs; remoteness increases costs and reduces competition; and without adequate institutions there are difficulties in contract enforcement.

For smallholder farmers to fully participate in markets they must also be able to meet both observable and unobservable transaction costs. The former group is made up of marketing costs such as transport, handling, packaging, and storage. The latter transaction costs include cost of information search, bargaining, screening, monitoring, coordination, and contract enforcement. Uncertainty about these raises the risks to farmers, which have to be managed. Or they are avoided by opting out of markets.

Constraints that limit market participation by smallholder farmers include supply-side constraints (especially those at the farm level that limit sufficient and reliable flow of products), demand-side constraints that limit the growth of local consumption of agricultural products, and markets and marketing institutions that are not well linked to serve farmers, especially in rural areas.

Source: author

Linking farmers to markets is necessary, therefore, although there are debates about which markets are most appropriate and viable for small farmers to engage with, and how formal institutions can be structured to benefit the least powerful participants. The Via Campesina movement represents at least 200 million farmers and rural workers, plus a range of organizations and indigenous groups, worldwide (Rosset and Martínez-Torres, 2012). Via Campesina by no means recommends polices of autarky (subsistence outside commercial market exchange systems) but acknowledges the potential for smallholders to improve their well-being, food security, and self-esteem, and to forge an adequate livelihood, by engaging in local markets and avoiding engagement in oligopolistic global markets. 'Food sovereignty' is a development approach that aspires to equitable market participation by smallholders in the development of local food systems (Poole et al., 2013b).

It will help development strategies if we know the way and the extent to which farmers are linked to markets. Indicators of smallholder participation in output marketing would be:

- changes in producer/beneficiary incomes;
- changes in the distribution of income, for example between men and women;
- changes in total volume of product sales;
- proportion of product sales through new market outlets, compared with sales through 'traditional' or alternative outlets;
- product prices received through new market outlets, compared with 'traditional' outlets;
- proportion and levels of income from new economic activities, compared with 'traditional' activities.

The availability of such data on most rural economies is limited or non-existent:

> Ideally there would be statistics on degrees of commercialisation and average farm incomes by household, with similar statistics on other determinants of income ... However, no such comprehensive data sets exist, although some household surveys and other observations through time give indications of the relationships (Wiggins and Keats, 2013: 12).

What data are available will be related to the monitoring and evaluation activities of specific interventions and are generally not easy to find.

Monitoring markets reveals important dynamic patterns in production and consumption. For example, diets in many countries are shifting towards higher-value livestock products, fruits, and vegetables. Specific foods are often perceived to have significant health or other prized attributes. As some people grow in wealth and learn about nutrition and new products, demand expands and prices rise, offering new supply opportunities to farmers for commercial sales. At the same time, food is diverted from poorer consumers – and poorer rural people who are net consumers – who can no longer afford the product. Rural people who cannot produce enough food for their own

> **Box 1.5 Market dynamics and new opportunities: The case of Ethiopia**
>
> Ethiopia is one of the world's poorest countries, well-known for its precarious food security situation. But it is also the native home of teff, a highly nutritious ancient grain increasingly finding its way into health-food shops and supermarkets in Europe and America.
>
> Grown by an estimated 6.3 million farmers, fields of the crop cover more than 20% of all land under cultivation ... As Western consumers acquire a taste for teff, how to ensure that Ethiopia and its farmers benefit from new global markets is a critical question. Growing demand for so-called ancient grains has not always been a straightforward win for poor communities. In Bolivia and Peru, reports of rising incomes owing to the now-global quinoa trade have come alongside those of malnutrition and conflicts over land as farmers sell their entire crop to meet western demand.
>
> Ethiopia's growing middle class is also pushing up demand for teff, and rising domestic prices over the past decade have put the grain out of reach of the poorest. Today, most small farmers sell the bulk of what they grow to consumers in the city.
>
> This may have helped boost incomes in some rural areas but it has had nutritional consequences, says the government, as teff is the most nutritionally valuable grain in the country. Estimates suggest that while those in urban areas eat up to 61 kg of teff a year, in rural areas, the figure is 20 kg. The type consumed differs too: the wealthy almost exclusively eat the more expensive magna and white teff varieties; less well-off consumers tend to eat less-valuable red and mixed teff, and more than half combine it with cheaper cereals such as sorghum and maize.
>
> The government's agricultural transformation agency aims to boost yields by developing improved varieties of the grain, along with new planting techniques and tools to reduce post-harvest losses.
>
> David Hallam, former Director of the Trade and Markets Division at the UN's Food and Agriculture Organization, says that while there is money to be made from new global markets for traditional crops, governments have to support small-scale producers to ensure they share the benefits of increased trade.
>
> 'Typically, these products are going to go through many hands before they reach the shelves of Sainsbury's or wherever. There are [profit] margins at every step, and small farmers are not necessarily well placed to bargain with the bigger traders', says Hallam, who sees quinoa's popularity as a cautionary tale of how export opportunities can be a mixed blessing for poor countries.
>
> Regassa Feyissa, an Ethiopian agricultural scientist and former head of the national Institute for Biodiversity, warns that without careful planning, increased teff production for export may displace other important crops for farmers. And efforts to boost production could benefit business interests at the expense of small farmers.
>
> *Source*: Provost and Jobson (2014)

subsistence and can no longer afford to buy may have to leave the land, and farming becomes consolidated among less numerous but larger-scale producers.

Teff (or tef), the staple grain of Ethiopia, is one such 'health' food. A report by Minten et al. (2013) comments on the changing market characteristics in Ethiopia. Interventions in such situations can boost the rural economy, and sometimes even support the struggling smaller farmers. The authors outline ways in which to boost the potential for the teff economy by improving inputs, technology, and product marketing. Box 1.5 summarizes the new opportunities in such agricultural markets.

Engagement in dynamic markets is a necessary step for many smallholder farmers. The value chain concept developed by Kaplinsky and Morris (2002), among others, describes the sorts of opportunities and processes of which commercializing farmers can take advantage, referring to different types of product and market innovation that can lead to upgrading:

- *Process upgrading.* Increasing the efficiency of internal processes within and between individual links.
- *Product upgrading.* Improving old products or introducing new products through individual strategies or through interfirm processes.
- *Functional upgrading.* Adding value by changing the mix of activities and by assuming new functional roles and responsibilities.
- *Chain upgrading.* Investment in new but related enterprises, or diversification into unrelated enterprises.

Market access and constraints

Much research has been conducted on the attributes of smallholder farmers and the conditions required to enable smallholders to upgrade and diversify their productive activities and link with markets. Amrouk et al. (2013) consider that there are three sets of factors which condition smallholder market access:

- *Farm and farmer characteristics.* The level of education and resource endowments, the level of technology, land size, and quality, and the stock of other productive assets. They also include household structure, consumption needs, and risks faced, which make up the vulnerability context.
- *External factors.* For example, the prevailing physical and institutional infrastructure such as roads, electricity, communications, market, and rules of law, which drive price incentives and the decision to invest in technology and generate surpluses.
- *Macro and sectoral policies.* Policies affect market access through their impact on prices and trade incentives.

A wide range of factors affects the extent to which smallholder farmers can integrate production with commercial markets, as can be seen in Box 1.6.

Household heterogeneity

Due recognition has recently been given to the differences between households, and warning against development approaches that fail to differentiate adequately between different household types. The World Bank's adoption of a 'commercialization narrative' for smallholder agriculture in Africa has a 'big picture' approach to rural policy formulation. This tends to gloss over inter-household heterogeneity, to say nothing of intra-household heterogeneity, that is, the differences within households among men and women,

> **Box 1.6 Factors affecting smallholder market participation**
>
> Not all farmers can take advantage of market developments. Smallholder farmers' access to evolving agricultural markets – especially to value chains – is commonly constrained by diverse factors:
>
> - Smallholders face high costs due to geographic barriers such as remoteness or biophysical limits to productivity (e.g. water availability).
> - They have limited productive asset holdings – of land, livestock, labour, critical equipment – which may constrain the scale of production and limit the marketable surplus.
> - There are complex and variable institutional arrangements including conditions of contracts, product grades and standards, certification, and access to credit, insurance, and technical information through extension services and collaborative initiatives.
> - Traders tend to work with a limited number of larger 'preferred suppliers' who are able to guarantee a large and continuous supply of produce and more easily meet market specifications.
> - Smallholders have a propensity to avoid risk:
>   - Risks result from adverse weather, pests and diseases, volatile prices, volatile policy environments, which are disproportionately high due to difficulties faced in accessing market information, credit and other inputs, and technical assistance, depending on contractual arrangements and whether they participate in formal or informal markets.
>   - Higher-value markets can be subject to boom-bust cycles that swallow sunk costs and investments.
>   - Credit provision generates cash flow risks.
>   - The adoption of complex technology generates risks associated with production and therefore delivery.
>   - Contractual arrangements generate risk of malfeasance (criminal behaviour).
>
> *Source*: summarized from Arias et al. (2013: 21–2)

old and young, high and low status. Rural heterogeneity is understood best within the 'small picture' of rural household characteristics (Poole et al., 2013b). The importance of analysing the small picture – detail, disaggregation, contextual 'locality', and 'particularity' – is highlighted by Poole et al.

In its seminal *World Development Report* (WDR) on agriculture, the World Bank noted that 'heterogeneity defines the rural world' (World Bank, 2007: 5). Box 1.7 contests this view.

It is not a new thing to stress the importance of disaggregating the different characteristics of large populations and recognizing diversity. Barrett (2008) comments on diversity in African smallholder agriculture. As already noted, most smallholder farm production is for subsistence, but even households which sell staple grains – typically soon after harvest – are net consumers over the year, relying on income from the sale of cash crops or labour. Nevertheless, there is considerable differentiation and inequality among smallholder farmers. It is common for there to be a high level of seller concentration in staple-food product markets which tend to be dominated by the farmers who are better-endowed in terms of productive assets, technology,

> **Box 1.7 How much differentiation?**
>
> '... the level of "differentiation" in the WDR and in the mainstream literature is both limited and reductionist. It glosses over the development "losers", whose limited assets and capabilities consign them to exit from agriculture and often from rural life into – probably the lowest – echelons of an urban-industrial society. Exit from agriculture can mean unemployment, social disruption, and urban deprivation within a context of burgeoning populations, climate change, and resource scarcities.
>
> 'Thus, the levels of differentiation commonly used are not very "local" or "particular", reflecting the methodologies of meta-analytical approaches and the growing influence of thematic reviews. They do not get deep into the hearts and minds of rural household members. Differentiation and customization are conceived only within the overarching imperative of commercializing agriculture.'
>
> *Source*: Poole et al. (2013b)

and geographic location. Barrett's review showed that access to financial services is also associated with the level of market sales: use of credit and insurance is strongly and positively associated with wealth. This may be because 'liquidity permits households to invest in higher-yielding, longer cycle crops, in seasonal inputs that boost yields, and in improved production technologies that require some initial sunk costs ... [there appear to be] multiple pathways through which private wealth affects market participation' (Barrett, 2008: 310).

Drawing on Barrett's work, Arias et al. (2013: 10–11) likewise emphasize the heterogeneous nature of households and transaction characteristics under three headings of household assets, market connectivity, and functionality. Smallholder farmers and their contexts differ according to the following:

- The smallholder household's access to, and the productivity of, assets, including natural resources, labour, and capital, vis-à-vis their subsistence needs will determine both their ability and their willingness to increase production for sale in markets.
- The connectivity of smallholders to different markets, which can be considered in terms of remoteness (defined broadly to include geographical proximity, knowledge asymmetries, power relationships, and the costs of commerce, or 'transaction costs'), will modify the incentives that they receive.
- The functionality of these markets. Many local food markets are volatile because of the low volumes transacted and their limited integration with regional or international markets, which limits the market's ability to modify demand- and/or supply-side shocks. Volatility can affect the level and riskiness of returns to the producer. Where markets are not well integrated, returns to increased output can diminish quickly as prices plummet, significantly affecting incentives for market participation and, consequently, for productivity-enhancing technology adoption.

Following Sen's articulation of entitlements, freedoms, and capabilities (Sen, 1981), the livelihood capitals framework focuses attention on human capacities and constraints, and on exploring smallholder heterogeneity. There is much more to be drawn from the literature on livelihoods and market participation, which will follow in Chapter 2.

Inequality and opportunity: market exclusion and inclusion

The corollary of heterogeneity is inequality, and it is important to include ethical implications of heterogeneity. How these are interpreted is a philosophical question, but most non-philosophers will have opinions on ethical questions. Inequality is 'unevenness', a feature of heterogeneity, and in itself is not necessarily 'bad', but more a fact of life. But one thing leads to another: inequality leads to inequity – which is 'unfairness' and, in most ethical frameworks, is considered to be bad; and inequity ultimately leads to iniquity – which is 'wickedness' (Poole, 2005). Almost everyone would agree that iniquity is a synonym for bad, not only bad for individuals but bad for society more widely, not least because of the probability of social and political discord.

Increasing awareness of inequality and its consequences is why one of the SDGs is a specific goal to 'Reduce inequality' (Goal 10, located in the second level of Figure 1.1). More specifically, SDG 5 is to 'Reduce gender inequality'. And without greater equality, it will be difficult to achieve other goals, particularly SDG 16, 'Promote peaceful and inclusive societies, justice and accountability'. Development practitioners may hesitate to enter debates about philosophy, but the ethical dimensions of development strategies, policies, and practice are firmly entrenched within the Sustainable Development Goals.

Two examples highlight the significance of inequality in the development discourse, and many others would serve to emphasize the point:

- According to Berdegué and Fuentealba (2011), in Latin America during the early 1980s there were 124 million rural inhabitants, 74 million of whom were poor, and of whom 41 million were hungry; after three decades, the numbers are 119 million, 62 million and 35 million, respectively. Meanwhile, in the same period GDP per capita increased by over 25 per cent in real terms.
- Like parts of Latin America, India has experienced dramatic economic growth in recent decades. But India's growth has been concentrated among a growing middle class that is still hugely outweighed by the mass of the population left in poverty. A recent publication has commented that the average annual GDP growth of 4.2 per cent between 1990 and 2005 has failed to reduce the burden of poverty and insecurity at a rate comparable to most developing countries – this is the Indian enigma (Pritchard et al., 2014).

While extolling the rates of development in the global South, the United Nations Human Development Report also draws a distinction between economic growth and human development progress. The middle class in the South is growing in size, income, and expectations, but inequality is high. Not all countries and not all people within a country benefit from the patterns of Southern growth. The Report draws particular attention to the impact of education policies, the failure of which 'will adversely affect many essential pillars of human development for future generations' (United Nations, 2013).

Social and economic inclusion as a means to overcome inequality are important and have been much debated. Helmsing and Vellema (2011) draw attention to the importance of poor people's 'voice', which enables the poorest to assess the advantages of poverty reduction interventions against the trade-offs and risks. While inequality is assumed to be undesirable, 'inclusion' as an objective itself needs further examination. For Helmsing and Vellema (2011: 12), 'inclusion and exclusion are more usefully seen as processes shaping how (rather than if) actors participate'. They note that some actors may be excluded or 'self-exclude' from a particular economic enterprise in favour of other activities and networks.

Thus, identifying and targeting 'beneficiaries' of market initiatives and the scope for upscaling successful interventions become important policy dilemmas; and voice should be given to the 'targets' so that they can consider strategic alternatives. Summing up the introduction to their edited volume on poor people's inclusion in commercial agribusiness chains, Helmsing and Vellema (2011: 18) comment:

> The development impacts of inclusion of small producers, local firms and workers in (global) value chains importantly depends on two conditions: the terms of participation in the process of inclusion and the degree of alignment of value chain logics with the capacities of actors and the institutions.

That people should be central to the processes affecting their lives now seems obvious. Nevertheless, policymakers too easily make statements that abstract from the complications of reality. For example, the Government of Uganda signalled the importance of arable agriculture in promoting the food security of the pastoralist Karamojong people.

> 'Karamoja has for a long time been buying food from her neighbours, this is the time to change the trend so that Karamoja's neighbours can rely on her for food', said the [Karamoja Affairs] Minister after she was impressed by the many acres of different foods grown ... [The Minister] commended this work saying that if many families embraced farming on a large scale, Karamoja would get to the desired level of development. She saluted the families who have tirelessly worked in the fields to grow food to feed their families and urged them

to produce more so that they can remain food secure throughout the year and even have surplus for sale (ChimpReports, 2013).

However, the plan to make every household cultivate arable crops in a garden has been criticized by many development authorities for failing to consider the social, economic, and agroecological contexts and involving the people themselves in determining their livelihood strategies (IRIN, 2014).

Thus, voice and choice both matter in identifying and prioritizing problems, and in the planning and implementation of agricultural development and market participation policies. Credence has to be given to the complexities of household management: smallholders will not respond automatically to initiatives to promote agricultural commercialization. As well as policy incentives, the local context, individual characteristics, and the opportunities and threats of the external environment affect smallholder choices:

> Expanding commercial agriculture requires a decided mind-set: a commitment to farming and new technologies, a low threshold of risk aversion, willingness to invest in land and soils, access to finance, skills in managing business relations, price negotiation, time spent in markets, product and process quality control and assurance, continuous improvement and efficiency enhancement. Also, collective activity with the inherent operational and management challenges is necessary for minimizing transaction and transformation costs, and involves significant individual and organizational learning. This commercializing metanarrative is much riskier than subsistence farming, rural wage labour and/or migration, and may not be an attractive profession (Poole et al., 2013b: 164).

The limitations of agricultural development for poverty reduction

A critical issue underlying the research and knowledge reported in this book is the commensurability between development interventions and objectives. By this we mean that initiatives and interventions should address appropriate levels of objectives and not pretend to cure all the market problems noted above. It has been said that 'The same objective can often be served by several alternative policy instruments; and the same policy instruments can affect the attainment of several policy objectives' (ILRI, 1995). This is largely true, and therefore it is necessary to be wary of the 'single instrument trap' which posits that 'there is only one way (instrument) to tackle a problem or set an objective' (ILRI, 1995).

At the same time, commensurability between an instrument and an objective is a warning against unreasonable expectations that a 'low-level' instrument can achieve a 'high-level' objective. This concern that expectations should be realistic and grounded in reality is certainly applicable to smallholders' access to markets: as we have noted, improving smallholders' market access is an objective per se, and may contribute to higher-level objectives such as poverty reduction. The needs of the poorest smallholders 'are probably best met by creating jobs, building their assets, improving their health and education, and in providing social protection. Market links will not provide multiple wins' (Wiggins and Keats, 2013: ix).

The complexity and interrelatedness of the new SDGs has been referred to already: the potential synergies and conflicts between the goals will create numerous policy dilemmas and require careful examination of the outcomes of interventions across sectors in ways to which development specialists are not accustomed. For example, boosting agricultural growth may adversely affect resource management and climate change; promoting female participation in the agricultural economy may adversely affect girls' education and household caring practices. Many other conflicts and trade-offs are plausible, giving rise to considerable governance challenges (Waage et al., 2015).

What then are the scope and limitations of improving smallholders' access to markets, of agricultural development? What also are the necessary accompanying approaches? What we have is a tension between two propositions. First, we agree that there are multiplier effects of agricultural development on broader processes; one strategy can have wide impacts, as shown in Figure 1.3 which makes explicit the connections between agricultural development and wider development processes.

Second, we face the reality that addressing even one high-level goal requires multiple approaches. Agricultural development depends on a range of other factors (Figure 1.4).

If addressing such goals requires multiple and integrated strategies, aspirations for what can be achieved by increasing smallholder participation in agricultural market development must be limited, rather than resolving all development problems. Thus, the impact of specific and focused actions to improve smallholders' market access should not be judged by the wider impact on overall poverty reduction and other SDGs.

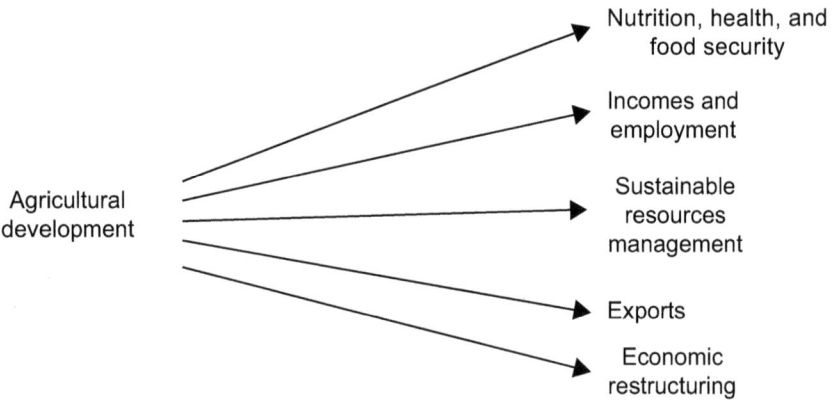

Figure 1.3 Agriculture's contribution to other sectors, activities, and policies

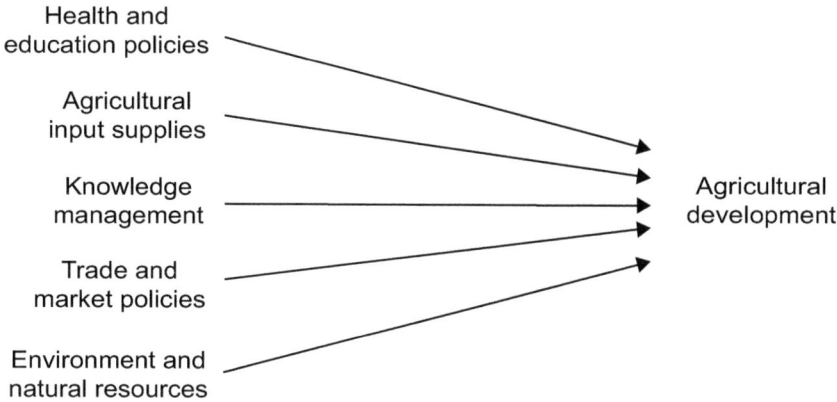

Figure 1.4 Agriculture's dependence on other sectors, activities, and policies

Here we focus on a more discrete and immediate objective within the processes of agricultural development. The task is to enable smallholder farmers to make sales (and perhaps purchases) of goods and services in markets which hitherto have been closed or non-existent. One of the secondary effects will be to boost the sector as a whole by raising incomes, creating employment, increasing foreign exchange earnings, substituting agricultural imports, and increasing food security. But this process alone cannot be expected to resolve the wider rural needs of reducing poverty and hunger, improving health and nutrition, promoting gender equality, facilitating economic restructuring, and stimulating industrial development.

Nevertheless, it is important not to narrow the analytical scope unduly. Important considerations are equity and sustainability: we must understand who accesses the markets and benefits from the sales, and whether the benefits are likely to last in the long term. Gender impacts, the interests of the poorest, and environmental sustainability are all relevant policy dimensions when considering smallholders' access to markets. The thinking and research covered in this book are concerned with this lower level in the hierarchy of policy formulation and development practice.

A note on terminology

We have noted that there are goals, trade-offs, conflicts, constraints, and priorities in the political processes that affect the expectations and achievements of policies and approaches to improving market access. A common understanding of these concepts will help the subsequent analyses and discussions. This is how these terms will be used:

- *Priorities*. With limited resources, there is usually a hierarchy or ranking whereby one objective is considered to be more important than another.

- *Trade-offs.* One objective has to be set aside in favour of another which is assigned higher priority.
- *Conflicts.* Achieving one objective reduces the possibility of achieving, or opportunity to achieve, another objective.
- *Constraints.* The existence of priorities, trade-offs, and conflicts imposes limits on what can be achieved, and together often imply minimum thresholds for action within the specific development context and the wider policy environment.

Decisions about priorities, acceptance of trade-offs and resolution of conflicts are normative, part of the political process, which in turn and ideally can be influenced by voters and stakeholders, particularly from among beneficiaries and advocacy organizations.

Moreover, we have already employed a variety of terms that are used to discuss development issues and policy, including goal, objective, target, indicator, approach, mechanism, and instrument. When thinking about agricultural development, greater precision is helpful. For clarity of communication and understanding, from here on the intention is to refer to the following:

- *Objective.* A desired state that can be, inter alia, human, natural, economic, or technological and can range within a hierarchy from the 'low level', specific, and small scale to 'high level', large in scale and scope.
- *Goal.* A high-level objective, usually long term in nature and large in scale and scope.
- *Target.* A low- or intermediate-level objective that may be short to medium term and which contributes to achieving a goal.
- *Indicator.* A measurable phenomenon enabling assessment of progress towards reaching a target.
- *Policy.* A defined and coherent programme formulated to achieve an objective, usually medium to long term, with a set of specified actions and activities.
- *Approach.* The conceptual development and empirical experience which together constitute a theory of change and the evidence base for a particular policy formulation.
- *Intervention.* A project, action, or activity of an agency (public, private, third sector) external to the targeted beneficiaries, undertaken as part of a policy formulated to achieve an objective.
- *Initiative.* A project, action, or activity which arises from within or among beneficiary organizations and individuals.
- *Instrument.* The specific mechanism, means, and methods whereby an intervention or initiative is implemented.

These definitions cascade downwards from the general to the particular. Similarly, the results of specific interventions and initiatives are building blocks on which higher-level policies are formulated and higher-level objectives can be achieved, which is a process of escalation upwards: a 'policy pyramid'. While this is intuitively plausible, the mechanisms – or a theory of change – through which the upwards escalation of impacts and effects are, or should be, mediated are usually not clearly articulated. It is important to make these connections in order to understand how escalation of local development outcomes may (fail to) contribute to higher level objectives.

About the book

It will become clear that this book is about the success or otherwise of efforts at addressing the low-level objective of market access and participation. The dimensions of market access – access by whom and for how long – fall within the analytical scope of the studies reported but at the same time the objective of the book is itself limited. Considering the hierarchy of objectives to which we have referred, improving smallholders' market access is an intermediate-level objective. Positive impacts on higher-level poverty reduction objectives can be expected but are attenuated by these realities. Therefore, the extent to which the research and development activities reported here achieve their objectives should not be measured by the many higher-level indicators that together reflect the wide range of policies and objectives subsumed within the SDGs, concepts of sustainable development, and comprehensive formulations of economics, ethics, and the environment. In short, improving market access alone cannot solve rural poverty.

Nevertheless, the book will address two questions drawing on concepts and empirical results of a programme of work undertaken by the United Nations Food and Agriculture Organization (UN FAO):

- it will indicate the strengths and weaknesses of interventions and initiatives formulated to improve smallholder market access; and
- considering the policy realities explained above, it will evaluate improving market access as an approach which contributes to bigger development goals.

From diverse international research and development activities and reports, this book distils new knowledge on models of agricultural development which create opportunities for small-scale agricultural entrepreneurs, and specifically for improving smallholders' access to markets. It tackles four conundrums:

- It recognizes that universal smallholder participation in commercial markets is unattainable and also that commercial elitism is undesirable.
- It accepts that there are economies of scale in agricultural production, but also asserts that diseconomies occur at both extremes: large scale and small scale.
- It rejects a model of pure, large-scale agricultural capitalism which squeezes smaller players out of the market and too often actually off the land which is the basis of their livelihood, patrimony, and social meaning.

- It fosters the notion of inclusive forms of agricultural entrepreneurship, meaning the opportunities created for small-scale farmers to coordinate and integrate their production with larger-scale agribusiness.

The implication for policies of agricultural development, export, and food security is that it is important not to promote large-scale agricultural capitalism and agribusiness at the expense of smallholder inclusion in economic development.

The book draws on a wide academic and development literature, including much of the author's own work. The core cases described in the book (Chapters 6–9) are those obtained through the All African, Caribbean and Pacific (ACP) Agricultural Commodities Programme (AAACP) funded by the European Union, which ran from 2008 to 2011 (see FAO, undated, All ACP Agricultural Commodities Programme).

The AAACP, with a budget of €45 m, was an inter-agency initiative implemented by FAO, the Common Fund for Commodities (CFC), the International Trade Centre (ITC), the United Nations Conference on Trade and Development (UNCTAD), and the World Bank.

The overall objective of the programme was to reduce income vulnerability and improve the livelihoods of producers dependent on agricultural commodities in the Africa, Caribbean and Pacific regions by building the capacity of actors along commodity chains to develop and implement sustainable value chain strategies. The development cases were principally managed by FAO, whose mandate is to improve nutrition, increase agricultural productivity, raise the standard of living in rural populations, and contribute to global economic growth (see FAO, undated, About FAO).

The case studies elaborated in Chapters 6–9 and further analysed in Chapter 10 were funded by the CFC and implemented by the FAO. They are used to highlight the interrelationship between projects and participants, interventions and beneficiaries, organizations and individuals. One important factor that conditions market development and the involvement of smallholders is the role of government. Recognition that markets are not often free and fair has promoted the view that appropriate regulation is important to boost economic growth and performance. This book will not delve into regulatory issues, important though they are. It suffices to end this introduction with the prevailing World Bank view of regulation in agribusiness, which hints at some of the issues such as market failure, transaction costs, and risk which feature in later chapters, but which the avid reader can follow up separately:

> It is crucial to have regulations that can lower risk by enabling farmers to operate in a context where the outcomes of their decisions are more predictable. Governments need to strike the right balance between correcting market failures through regulations and minimizing the costs that those regulations impose on economic agents. This balance is essential for agriculture, but it is also particularly challenging (World Bank, 2017: x).

References

Africa Progress Panel (2012). *Jobs, Justice and Equity: Seizing opportunities in times of global change*. Africa Progress Report 2012. Retrieved 28 March 2017, from http://www.africaprogresspanel.org/publications/policy-papers/africa-progress-report-2012/.

Amrouk, E.M., Poole, N.D., Mudungwe, N. and Muzvondiwa, E. (2013). *The Impact of Commodity Development Projects on Smallholders' Market Access in Developing Countries: Case studies of FAO/CFC Projects*. Rome, United Nations Food and Agriculture Organization. Retrieved 28 March 2017, from http://www.fao.org/docrep/017/aq290e/aq290e.pdf.

Angelucci, F., Cafiero, C. and Malua, M. (undated). *A Supply Chain Finance and Risk Management Scheme for Addressing Financial Needs and Constraints to Smallholders' Participation in Samoan Fruits and Vegetables Value Chains*. Rome, Food and Agriculture Organization of the United Nations.

Arias, P., Hallam, D., Krivonos, E. and Morrison, J. (2013). *Smallholder Integration in Changing Food Markets*. Rome, Food and Agriculture Organization of the United Nations. Retrieved 28 March 2017, from http://www.fao.org/docrep/018/i3292e/i3292e.pdf.

Barrett, C.B. (2008). Smallholder market participation: concepts and evidence from eastern and southern Africa. *Food Policy* **33**(4): 299–317 <http://dx.doi.org/10.1016/j.foodpol.2007.10.005>.

Berdegué, J.A. and Fuentealba, R. (2011). *Latin America: The state of smallholders in agriculture*. IFAD Conference on New Directions for Smallholder Agriculture. Rome, International Fund for Agricultural Development.

Breisinger, C., Ecker, O., Al Riffai, P. and Yu, B. (2012). *Beyond the Arab Awakening: Policies and investments for poverty reduction and food security*. Washington, DC, International Food Policy Research Institute. Retrieved 28 March 2017, from http://www.ifpri.org/sites/default/files/publications/pr25.pdf.

ChimpReports (2013). 'Janet moves to reverse Karamoja's worsening hunger'. Retrieved 22 May 2017, from http://www.chimpreports.com/14024-janet-moves-to-reverse-karamojas-worsening-hunger/.

Devaux, A., Torero, M., Donovan, J. and Horton, D., eds, (2016). *Innovation for Inclusive Value Chain Development: Successes and challenges*. Washington, DC, International Food Policy Research Institute (IFPRI).

FAO (2012). *Smallholders and Family Farmers*. Retrieved 28 March 2017, from http://www.fao.org/fileadmin/templates/nr/sustainability_pathways/docs/Factsheet_SMALLHOLDERS.pdf.

FAO (2013). *The State of Food and Agriculture 2013: Food systems for better nutrition*. Rome, Food and Agriculture Organization of the United Nations. Retrieved 28 March 2017, from http://www.fao.org/publications/sofa/2013/en/.

FAO (2016). *The State of Food and Agriculture 2016: Climate change, agriculture and food Security*. Rome, FAO. Retrieved 02 November 2016, from http://www.fao.org/3/a-i6030e.pdf.

FAO (undated). 'All ACP Agricultural Commodities Programme'. Retrieved 27 April 2017, from http://www.fao.org/economic/EST/AAACP/en/.

FAO (undated). 'About FAO'. Retrieved 27 April 2017, from http://www.fao.org/about/en/.

Helmsing, A.H.J. and Vellema, S., eds, (2011). *Value Chains, Social Inclusion and Economic Development: Contrasting theories and realities*. London, Routledge.

IFAD (2011). *Feeding future generations: Young rural people today – prosperous, productive farmers tomorrow. Proceedings of the Governing Council High-Level Panel and Side Events, in conjunction with the Thirty-fourth Session of IFAD's Governing Council*. Rome, United Nations International Fund for Agricultural Development (IFAD). Retrieved 28 March 2017, from http://www.ifad.org/events/gc/34/panels/proceedings.pdf.

IFAD (2016). *Rural Development Report 2016: Fostering inclusive rural transformation*. Rome, United Nations International Fund for Agricultural Development. Retrieved 28 March 2017, from https://www.ifad.org/ruraldevelopmentreport.

ILRI (1995). *Livestock Policy Analysis. ILRI Training Manual 2*. Nairobi, Kenya, International Livestock Research Institute (ILRI). Retrieved 22 May 2017, from http://www.fao.org/wairdocs/ilri/x5547e/x5547e0i.htm.

IRIN (2014). 'Alarm over mandatory gardens in Karamoja.' Retrieved 28 March 2017, from http://www.irinnews.org/report/99523/alarm-over-mandatory-gardens-in-karamoja.

Kaplinsky, R. and Morris, M. (2002). *A Handbook for Value Chain Research. Prepared for IDRC*. Retrieved 28 March 2017, from http://www.value-chains.org/dyn/bds/docs/395/Handbook%20for%20Value%20Chain%20Analysis.pdf.

Karembu, M. (2013). 'Preparing youth for high-tech agriculture'. In R.B. Heap and D.J. Bennett, eds, *Insights: Africa's future ... can biosciences contribute?* pp. 91–97. Cambridge, UK, Banson.

LANSA (undated). Leveraging Agriculture for Nutrition in South Asia. Retrieved 27 April 2017 from http://www.lansasouthasia.org/.

Maestre, M., Poole, N. and Henson, S. (2017). Assessing food value chain pathways, linkages and impacts for better nutrition of vulnerable groups. *Food Policy* **68**: 31–39 <http://dx.doi.org/10.1016/j.foodpol.2016.12.007>.

Minten, B., Tamru, S., Engida, E. and Kuma, T. (2013). *Ethiopia's Value Chains on the Move: The case of teff*. Washington, DC, International Food Policy Research Institute. Retrieved 28 March 2017, from https://lirias.kuleuven.be/bitstream/123456789/456070/1/ESSP+Working+Paper+52.pdf.

Poole, N.D. (2005). Poverty, inequality and ethnicity: a note to policy makers on Latin America. *Eurochoices* **4**(3): 44–49 <http://dx.doi.org/10.1111/j.1746-692X.2005.00017.x>.

Poole, N.D., Alvarez, F., Vazquez, R. and Penagos, N. (2013a). Education for all and for what? Life-skills and livelihoods in rural communities. *Journal of Agribusiness in Developing and Emerging Economies* **3**(1): 64–78 <http://dx.doi.org/10.1108/20440831311321656>.

Poole, N.D., Chitundu, M. and Msoni, R. (2013b). Commercialisation: a meta-approach for agricultural development among smallholder farmers in Africa? *Food Policy* **41**(August): 155–165 <http://dx.doi.org/10.1016/j.foodpol.2013.05.010>.

Poole, N., Audia, C., Kaboret, B. and Kent, R. (2016a). Tree products, food security and livelihoods: a household study of Burkina Faso.

Environmental Conservation **43**(4): 359–367 <http://dx.doi.org/10.1017/S0376892916000175>.
Poole, N., Echavez, C. and Rowland, D. (2016b). *Stakeholder perceptions of agriculture and nutrition policies and practice: evidence from Afghanistan*. LANSA Working Paper Number 9. Chennai, India, Leveraging Agriculture for Nutrition in South Asia (LANSA), MS Swaminathan Research Foundation. Retrieved 07 April 2017, from https://assets.publishing.service.gov.uk/media/5963972ee5274a0a59000157/Mapping_stakeholder_perceptions_Afghanistan_on_template_0.pdf.
Pritchard, B., Rammohan, A., Sekher, M., Parasuraman, S. and Choitani, C. (2014). *Feeding India: Livelihoods, entitlement and capabilities*. Abingdon, UK, Routledge.
Provost, C. and Jobson, E. (2014). Move over quinoa, Ethiopia's teff poised to be next big super grain. *The Guardian*. Retrieved 28 March 2017, from http://www.theguardian.com/global-development/2014/jan/23/quinoa-ethiopia-teff-super-grain.
Rogers, T.S., Morrison, J.A. and Bammann, H. (2010). *Agriculture for Growth: Learning from experience in the Pacific. Summary results of five country studies in Fiji, Samoa, Solomon Islands, Tonga and Vanuatu*. EU-AAACP Paper. Rome, Food and Agriculture Organization of the United Nations. Retrieved 28 March 2017, from http://www.fao.org/docrep/013/am011e/am011e00.pdf.
Rosset, P.M. and Martínez-Torres, M.E. (2012). Rural social movements and agroecology: context, theory, and process. *Ecology and Society* **17**(3): C7–17 <http://dx.doi.org/10.5751/ES-05000-170317>.
Sen, A. (1981). *Poverty and Famines: An essay on entitlements and deprivation*. Oxford, UK, Clarendon.
Sutherland, P.D. (2013). Migration is development: how migration matters to the post-2015 debate. *Migration and Development* **2**(2): 151–156 <http://dx.doi.org/10.1080/21632324.2013.817763>.
United Nations (2013). *Human Development Report 2013. The Rise of the South: Human progress in a diverse world*. New York, United Nations Development Programme. Retrieved 28 March 2017, from http://hdr.undp.org/en/2013-report.
UN-DESA (undated). UN Sustainable Development Knowledge Platform. United Nations Department of Economic and Social Affairs. Retrieved 27 April 2017 from https://sustainabledevelopment.un.org/index.php?menu=1300.
United Nations Department of Economic and Social Affairs (undated). UN Sustainable Development Knowledge Platform. Retrieved 27 April 2017 from https://sustainabledevelopment.un.org/index.php?menu=1300.
Waage, J. and Yap, C., eds, (2015). *Thinking Beyond Sectors for Sustainable Development*. London, Ubiquity Press.
Waage, J., Dangour, A.D., Hawkesworth, S., Johnston, D., Lock, K., Poole, N.D., Rushton, J. and Uauy, R. (2011). *Understanding and Improving the Relationship between Agriculture and Health*. A review commissioned as part of the UK Government's Foresight Project on Global Food and Farming Futures. London, The Government Office for Science. Retrieved 22 January 2016, from http://www.lcirah.ac.uk/_assets/Foresight%20Report%20Agriculture%20and%20Health%20review.PDF.

Waage, J., Yap, C., Bell, S., Levy, C., Mace, G., Pegram, T., Unterhalter, E., Dasandi, N., Hudson, D., Kock, R., Mayhew, S., Marx, C. and Poole, N. (2015). 'Governing Sustainable Development Goals: interactions, infrastructures, and institutions.' In J. Waage and C. Yap, eds, *Thinking Beyond Sectors for Sustainable Development*, pp. 79–88. London, Ubiquity Press.

Wiggins, S. and Keats, S. (2013). *Leaping and Learning: Linking smallholders to markets in Africa*. London, Agriculture for Impact, Imperial College and Overseas Development Institute.

World Bank (2007). *World Development Report 2008: Agriculture for development*. Washington, DC, World Bank. Retrieved 28 March 2017, from http://siteresources.worldbank.org/INTWDR2008/Resources/WDR_00_book.pdf.

World Bank (2016). World Development Indicators 2016. Retrieved 28 March 2017, from http://data.worldbank.org/products/wdi.

World Bank (2017). *Enabling the Business of Agriculture 2017*. Washington, DC, World Bank. Retrieved 28 March 2017, from http://eba.worldbank.org/~/media/WBG/AgriBusiness/Documents/Reports/2017/EBA2017-Report17.pdf.

CHAPTER 2
Policy approaches and theoretical considerations

This chapter discusses the evolution of policies and approaches to agricultural development in recent decades. We introduce theory and literature relevant to smallholder farming; understanding how households and individuals behave is key to stimulating people's own initiatives and in designing and implementing development policies to improve livelihoods. The costs and benefits of engaging in markets to boost development are an important determinant of behaviour, as are the assets and attributes that each individual and household can make use of. The chapter closes by introducing the concept of the value chain as a development paradigm for promoting markets and development, and the particular implications for participation of smallholder farmers.

Keywords: households, transaction costs, livelihoods, capital assets, value chains, participation

Changing policies: from market systems to value chains

The post-colonial period

For developing economies, the potential of marketing systems to assure the availability of foodstuffs to rural and urban populations and to stimulate development drove research in the early decades of the post-colonial era. Reflecting on the first two decades after the Second World War, Jones (1974) noted the finding, novel for many analysts, that marketing was not a new phenomenon, but was 'much more common in the pre-colonial period than was previously thought' (Jones, 1974: 4).

System inefficiencies were often attributed to market imperfections and exploitative traders, suggesting the need for significant interventions to improve market performance (Jones, 1974). For many developing countries, the balance of initiative in promoting agricultural marketing lay with the state rather than with the individual until the late 1980s and early 1990s. The expansion of public-sector marketing could not have been undertaken without the provision of aid from bilateral and multilateral donors, whose policies constituted an endorsement of a state-led strategy. Marketing boards and cooperatives were a convenient counterpart agency for donors whose programmes of food aid, infrastructural investment, and rural development projects were increasing in importance.

Market liberalization

The period of structural adjustment through fiscal austerity, privatization, and trade liberalization, and the Washington Consensus phases of international development policy, followed the early post-colonial agricultural development policy of state-led intervention. There were diverse patterns of market reform, from the redefinition of parastatal roles, through a range of radical changes, to abolition of state intervention. The nature of the policies being adopted depended partly on ideology, and partly on assumptions about the capacity of the private sector to respond. Other considerations were the estimate of risk of failure in unprofitable markets, the importance accorded to social or 'public' objectives such as stockholding and pan-territorial pricing, the existence of scale economies, and the danger of emerging privatized monopolies.

Reduction in the state management of price and other policies was accompanied by reduced provision of services such as agricultural inputs and extension (Poole et al., 2013b). The potential for a response from the private sector to economic incentives and opportunities should have created a sense of optimism in stimulating efficient and competitive markets, and overcoming the livelihood constraints of poor producers and traders. The propensity of producers to engage in trade should have been a sound foundation for market-led economic growth. However, the resurgence of faith in liberal market mechanisms towards the end of the 20th century was paralleled by international policy neglect of the rural economy, and an uninterested attitude to, even a disengagement from, agriculture.

As it turned out, liberalization policies did not generally lead to efficiency and market-led growth. As traders were not willing or able to fill the void, farmers were left without market outlets. The radical reduction in the scope of state intervention and the not infrequent collapse of organized marketing systems did not stimulate a strong private-sector response, nor generate higher levels of competition. On the contrary, liberalization tended to precipitate a decline in agricultural trading. Indeed, many interventions were counterproductive. Barrett has noted that 'Coupled with exchange rate devaluation or depreciation that drives up the cost of tradable inputs (e.g., fuel), many market-oriented reforms of the past twenty or so years have sharply increased the costs of commerce, driving some regions and households back towards subsistence production' (2008: 311).

Revising policies and approaches

In looking for explanations for this policy failure, the costs of transactions in agricultural marketing were increasingly recognized during the 1990s as an obstacle to market efficiency. At the end of the century it was clear that most farmers without commercial knowledge or experience were unable to engage successfully in marketing their produce on their own account, and the possibility for many smallholders of engaging in high-value export markets was remote. Therefore, following the promotion of more liberal market mechanisms and into the post-Washington Consensus, policymakers began to look for intervention mechanisms to overcome both market and state failures in order to reduce poverty through economic growth.

The context of agricultural marketing in the new millennium has also changed dramatically. Much more has been learned since then about how real markets actually function. The recognition then of the opportunity to create marketable surpluses, and to exchange goods and services over larger distances in order to make these foodstuffs available, anticipated the explosion of international trade together with the social, economic, and political processes of globalization with which we are now familiar. This was facilitated by liberal market economics, technological change, and expanding commercial strategies, and was affected by institutional intervention for (and against) international agricultural development and trade through the General Agreement on Tariffs and Trade (GATT) and World Trade Organization (WTO) processes.

It is evident that donor ideology has also influenced the choice of organizational forms and marketing interventions (Poole, 2010). Donor policies are in turn influenced by pragmatic considerations such as how aid can be most easily dispensed and impact monitored. This probably applies just as much at the beginning of the 21st century as during the 20th century, and the danger of interventions representing prevailing conventional wisdom remains. A wide range of considerations come into play for influencing policy choices: re-evaluation of the positive role of the state in facilitating the Green Revolution; a resurgence of interest in collective organizations; and the development of the stakeholder concept. Comparative approaches to political economy and network approaches to business are beginning to vindicate a hybrid policy approach, drawing on models of both 'liberal' and 'coordinated' market economies (Hall and Soskice, 2001). That is to say, while the private sector accounts for most successful economic activity, market failures are still pervasive in developing countries and require both initiatives from the private sector and interventions from the state and the 'third sector'. Critical voices increasingly question the value of external development aid when compared with financial resources which originate within developing countries. Capitalizing on local financial resources and business initiative can enable smallholders and other small enterprises to integrate into modern value chains. Where this can be done without external intervention it is simply 'good business' (Harper et al., 2015).

Making markets work for the poor

Since the early 2000s, considerable attention has been attached to improving the performance of the wider business environment. The approach of 'Making markets work for the poor' (MMW4P, or M4P) stresses the process of creating opportunities through increasing access to markets, achieving equitable and remunerative prices for goods and services, and reducing risk. M4P is a market-systems-development approach to poverty reduction adopted by a number of international organizations (Dorward and Poole 2004; The Springfield Centre 2008). In 2011 the UK Department for International Development (DFID) set out its intention to leverage private-sector initiatives for development and poverty reduction:

> A key tenet of DFID's approach to private-sector led growth and sustained poverty reduction is a concerted effort to promote accessible and well-functioning markets. Better functioning markets can increase the demand for, and supply of, more affordable and appropriate goods and services that better meet the needs of the poor as consumers. In turn, they also create opportunities and generate incomes for the poor as producers, employees and entrepreneurs (DFID 2013).

With the distinctive inclusion of patterns of consumption as well as production, this highlights the important contribution of the agri-food sector to both the quantity and quality of food supplies for health and nutrition.

The central idea of making markets work for poor people is that poor people's livelihoods are linked to market exchange. The focus is on interventions which address market-system imperfections. The market system is understood to be the arrangement of multiple players – business, government, and civil society – and the core, rules, and supporting functions through which economic exchange occurs, develops, adapts, and grows. It is said to be a construct through which both markets and basic services can be examined (The Springfield Centre, 2015).

The unit of analysis being the market system, the potential for replication and widespread roll-out of successful interventions is significant. In the M4P approach, interventions are designed based on an analysis of factors which relate closely to the project cycle:

- setting the strategic framework;
- understanding market systems;
- defining sustainable outcomes;
- facilitating systemic change;
- assessing change.

With a conceptual basis in economics, the strength of M4P is its focus on improving the commercial and policy environments which so strongly affect the attractiveness of 'doing agribusiness' (World Bank, 2016; World Bank Group, 2016). The World Bank's approach is motivated by the need for agricultural development to address the demands of a growing world population which might reach nine billion people by 2050. Analysis of regulatory practices enables stakeholders to identify the barriers that hinder the development of agribusiness and the transaction costs of dealing with government regulations. There is an inbuilt assumption that, given an enabling environment of adequate institutional structures and conduct, poor people will be able to benefit from engagement in such markets.

In diagnosing market imperfection and designing potential interventions, the M4P analysis is highly applied, and the intervention approach is somewhat mechanistic, inasmuch as the actual characterization and selection of target groups (diagnosis of 'the poor and their context') is rudimentary (see Box 2.1): market rather than household- or people-focused. It suggests a range of tools – 'socio-economic studies, census data, poverty assessments, livelihoods analysis, investment climate surveys, competitiveness analysis, drivers of change' (The Springfield Centre, 2008: 28) – but does not use the range of disciplinary approaches necessary to understand people's attitudes, aspirations, and vulnerabilities.

While M4P envisages commercial interventions and is 'pro-competitive structures', it is weak in the policy dimension and draws back from advocating innovative public-policy interventions and radical restructuring

---

Box 2.1 The M4P diagnostic process in practice

The poor and their context

- Facilitator's objectives are for large-scale, pro-poor growth in rural regions
- Rural area; hundreds of thousands of small farmers below poverty line; agriculture is main source of household income
- Rice is principal crop but some vegetable cultivation also: demand for vegetables significantly outstrips supply
- Declining crop yields recognized as key problem within the sector

Specific market system

- Vegetable value chain serves local or regional markets
- Farmers use a variety of inputs bought from local retailers
- Income from vegetable production is low due to low productivity
- Farmers lack knowledge about appropriate inputs and cultivation techniques
- Farmers get information from a variety of sources: other farmers, government extension workers, agricultural supply retailers, and NGOs
- Farmers are dissatisfied with information they receive from all these sources

Systemic constraints

- Farmers are most likely to turn to retailers for information as they are most widespread and accessible
- Retailers tend to push products to maximize their own modest income; they do not see themselves as a source of information
- Retailers buy their supplies from large-input suppliers, who also provide them with information and support, but only about their own products
- Input suppliers have the capacity and incentive to support retailers to become more effective sources of information: satisfied farmers are good for business

Intervention focus

- Changing the distribution-channel development practices of large input suppliers

Source: The Springfield Centre (2008: 35)

and resource reallocation. Linkages of enterprise concepts to movements advocating collective organization are also weak because there is little emphasis on market systems.

There are many lessons to be learned still about how to encourage smallholder inclusion in agri-food markets. On the one hand, poverty has been reduced through local and international economic growth in many contexts; on the other, the complexity of production systems, disadvantageous economic geography, bad governance, and climate change continue to assail and define the poorest populations. While there have been notable achievements in agricultural development and marketing over recent decades, food and agricultural marketing research have not provided widespread solutions to the current problems of global security and poverty reduction.

Among the inferences to be drawn from past experience, the most important is that ideological shifts have not resulted in viable policies to boost the contribution of agricultural marketing to poverty and hunger reduction in many situations. Because poverty problems are specific to a given development context, the demand for 'generalizability' and 'upscaling' of interventions is often ineffective and misplaced.

Consequently, it is necessary to emphasize empirical approaches to policy formulation based on precise and disaggregated studies. Effective interventions are likely to be local and particular, and therefore more costly than the ranges of rates of return and net present values to which economists are accustomed. On the contrary, targeting the local economy through interventions and investments in local assets, structures, and institutions is more likely to bring significant benefits to communities of the poor and hungry.

Institutional innovation

Agreeing with Barrett's view that 'The primary theme in the literature on smallholder market participation is the importance of transactions costs' (2008: 310), Poole and de Frece (2010) suggest that often it is small-scale institutional innovations in local market organization that serve best to stimulate smallholder participation in input and output markets: 'institutional innovation' is needed in respect of new 'rules of the game' and also new types of organization, i.e. 'new players in the game'. So, too, are specific investments in human and social capital, and business and market organization: new ways of organizing both people and markets to work for the poor.

It is interesting that, after decades of underperformance, there is a resurgence of interest in farmer organizations. Box 2.2 outlines the arguments in favour of collective organizations.

Challenged by the Millennium Development Goals, policymakers turned to value chain approaches, the delivery of specific business services, and the facilitation of wider enabling environments that might make market systems and chains work better for the poorest (Poole, 2010). Value chains will still

> **Box 2.2 Collective organization: Can it contribute to more equitable and efficient markets?**
>
> 'There are theoretical explanations of the failures of collective organisation but, at the same time, the fundamental reasons for collaborating hold true: the potential for exploiting production and managerial economies of scale, overcoming market entry barriers, reducing transaction costs, and cultivating supply chain relationships. Collective decision making may be cumbersome, and top-down decision making may be undesirable. But new forms of collective enterprise illustrate that innovative business models can work: "new generation cooperatives" may provide solutions to some of the historical and structural problems of cooperatives. There are alternative management structures and financial resources – either philanthropic support; or external equity investment with a capacity to exert leverage through management building; or invitations to bondholders with a financial stake but without governance rights. These strategies offer the possibility of external capitalisation without diluting membership control.
>
> 'Despite a history of operational failure, statutory arrangements such as levies and marketing boards are also mechanisms with potential to overcome market failures and the provision of public goods …
>
> '… The approach of external supporting organisations must be patient and realistic. Collective enterprise may not always work because usually there are threshold levels of asset requirements and of external support for successful group formation and operation. It is clear that collective enterprises are "organic": they learn and grow, sometimes fail, and sometimes need to rise from the ashes of incompetence and corruption. The path to maturity is usually long, and needs supportive investment through a range of planned and sequenced business services, with an exit strategy emplaced to ensure progress towards sustainability. And there is no "one size-fits all", and no guarantee that individual successes can be upscaled and replicated …
>
> *'What part does the institutional framework play?*
>
> 'The importance of the historical, political and market context within which smallholder enterprise operates is clear. Institutions frame the relationship between state and the "citizen" and "organisations", mediating the flows of technical support towards the grassroots, and advocacy towards the state, and the relationship of politics to local development processes. The purpose of the formal legal and regulatory framework, such as competition and business laws and cooperative laws is, in part, to shape the environment and enable business to operate effectively. This may or may not happen in practice: producer organisations often are surrounded by legal restrictions, and micro-, small and medium-sized enterprises often go unrecognised by the state as policy stakeholders. Where sectoral policy is increasingly directed towards scale, efficiency and new technologies to address food security objectives for growing populations in an era of climate change and social transformations, the needs of smallholders may be unrecognised and underprovided. Weaknesses in transport systems and infrastructure, and certain restrictive trade practices within and between African countries, also impose heavy burdens on local or regional trade. Such formal business and legal frameworks, policies, and priorities are critical to economic empowerment of the rural poor.'
>
> *Source:* Poole and de Frece (2010: 8–10).

be an important concept in the new Sustainable Development Goal era, not least Goal 8: 'Promote sustained, inclusive, and sustainable economic growth, full and productive employment and decent work for all' and Goal 12: 'Ensure sustainable consumption and production patterns'. We will revisit value chains later in this chapter.

Theoretical approaches to households, markets, and marketing for the poor

Households and economic decisions

Barrett (2008) traces the imperative for people to engage in market exchange to the advantages of specialization and gains consistent with comparative advantage explained by the classical economics of Adam Smith and David Ricardo. Such an approach has important insights but makes simplifying assumptions about the resources and technologies available to economic agents, and levels of information and patterns of decision making, that are difficult to uphold for many people engaged in modern markets, let alone smallholder farmers. In particular, the 'what?', who?', 'how?' and 'why?' decisions made in households about production and marketing, and the distribution of 'gains' from collective enterprise and within households, are complex management phenomena. Farmers' decision-making processes are dynamic, affected by a wide range of endogenous and exogenous factors, idiosyncratic, and difficult to model. There is no substitute for in-depth research to understand these phenomena, and there are serious problems with universalizing findings, implications, and policy proposals.

Nevertheless, theoretical economic constructs like those used by Smith and Ricardo can provide some partial but useful insights. But as Haddad et al. say, 'Although it is widely recognized that the welfare of an individual is, in large part, based on a complex set of economic and social interactions, development policies do not always acknowledge these' (Haddad et al., 1997: 1). A variety of economic modelling approaches have been developed to handle the complexities of intra-household decision making. Latterly, intra-household resource management and allocation has come to be viewed as both a process and an outcome (Bennett, 2013). These household models illuminate likely household power, gender, and intergenerational relations.

Modelling insights

In the theoretical understanding of household behaviour, there is firstly a unitary model whereby the household is considered to be a single decision-making unit led by a dominant and sometimes benevolent male (Becker, 1976). This is associated with the work of Becker and Mincer in the 1960s and 1970s, is consistent with the concept of a single welfare function, and gave rise to the New Home Economics. In the unitary model, economic efficiency concerns revolve around maximizing benefits for the household overall, by making investments with the best returns. Equity issues are those which affect, for example, distribution of benefits among children: 'First, parents may be interested in ensuring that all children are equally well off. Alternatively, they may have preferences for particular children; for example, boys over girls, firstborn over latter born, their own children over those whom they raise as foster children' (Haddad et al., 1997: 4).

Substantial criticism of this approach led to the development of alternative approaches which assume collective and/or consensual decision making by individuals within a multi-person household, and bargaining approaches, which assume non-cooperative behaviour. An alternative approach is that of independent-individual models of decision making (Grossbard, 2010), which can fall within a spectrum from pure individualism to pure cooperation among multiple players with one 'blended utility function and completely pooled resources' (Grossbard, 2010: 4). In Haddad et al. (1997) recognition is given to both quantitative and qualitative analytical approaches to household welfare, and also the need for economics to merge with other relevant disciplines, specifically anthropology and nutrition.

Bargaining models assume that individuals in households act either cooperatively or non-cooperatively. Other models allow for joint decisions or accept that members react to each other in certain ways, by cooperating in some spheres but acting independently in others. These features are illustrated by some recent analyses of age and gender relations:

- Female participation in decision making – or women's bargaining power – is commonly held to be influenced by level of education, incomes, and assets (Doss, 2013), but also has to do with the presence of parents in the household (Bayudan-Dacuycuy, 2013).
- Dauphin et al. (2011) found evidence for 'children' (aged between 16 and 21, and daughters, irrespective of their age) being household decision makers in a collective modelling approach to analysing data from the British Family Expenditure Survey. Evidence was inconclusive for sons and children over the age of 22.
- Lundberg et al. (2009) also found evidence that children's decision-making power was a response to a demand for autonomy as well as a function of parents' discretion and investment in child development.
- In terms of resource distribution, it can be inferred from work in China that having a first-born son conferred greater decision-making power on a woman than having a first-born daughter, resulting in improvements in the mothers' nutrition and reducing the probability of her being underweight (Li and Wu 2011).

Households and management decisions

Belief in intra-household heterogeneity and gender roles in decision-making patterns, and a concern for equity underline policies to target women in microfinance and social welfare programmes such as conditional cash transfers, and even land-tenancy arrangements. But as modelling approaches suggest, women are not isolated decision makers, even in situations of significant male outmigration, and in many cases decisions about resources are likely to be negotiated (Tincani, 2012).

Work in Burkina Faso, which may well have a wider resonance, shows there are various features of Sahelian societies which raise questions about patterns and changes in intra-household decision making. Firstly, the traditional pattern of polygamous and complex households (*la concession*) raises questions about the dominant role of men, or more specifically the male household head, vis-à-vis the adult women in the household. In terms of negotiating power, a ranking of adult women, or more specifically of wives of the male household head, can be expected to affect distributional outcomes for sub-households (*le ménage*) concerning access to and quality of land for agriculture (principally staple grain production), and rights to the grain store of the male household head.

Within the Sahel, different household endowments in terms of social and human capital, notably labour, land quantity and quality, and rainfall, are likely to give rise to different patterns of entitlements and decision making. The influences arising from the external environment are likely to be highly significant. The following changes are likely to impact rural household management dynamics:

- geography and infrastructure;
- level of services and market competition;
- opportunities for employment in the gold industry;
- migration to Côte d'Ivoire;
- the encroachment of Western market ethics and evolution of gender roles;
- traditional and Islamic social norms;
- ecological changes apparently as a consequence of climate change;
- state and private policies and investments and institutions (regulations, standards, taxation, and ethics);
- political instability in neighbouring countries.

Households and market decisions

Participation in agricultural input and product markets is one such complex decision-making space. Adding in the omnipresent policy interventions and development initiatives, 'opting out' of markets can still be economically rational for poor people (Helmsing and Vellema, 2011) because of idiosyncratic household preferences which are shaped, inter alia, by risk and vulnerability, and individual attributes and aspirations within this complex environment. Barrett (2008: 300) states that 'One thus has to get institutions and endowments, as well as prices, "right" in order to induce market-based development'. Assets, infrastructure, and incentives are preconditions for helping poor households out of a low-level equilibrium of 'semi-subsistence production by smallholders operating rudimentary production technologies with limited assets and participating modestly, if at all, in competitive and regionally or globally integrated markets offering remunerative terms of trade' (Barrett, 2008: 300). But in many parts of sub-Saharan Africa, price and trade interventions have not been sufficient to allow or encourage smallholders to enter staple food markets. Household heterogeneity and the contextual differences already mentioned generate transaction and other costs, plus risk, such that idiosyncratic obstacles are superimposed on the systemic challenges of accessing markets.

Reviewing empirical studies on African smallholder market participation from the 1990s onwards, Barrett (2008) identified a significant literature on access to high-value markets in a range of countries, and a growing literature on contract farming and linkages to supermarket retail systems, again, primarily for high-value products such as those from horticulture. While the lessons from Barrett's magisterial review are likely to have implications for a wider population, Donovan and Poole (2008) reviewed experiences of smallholder access to non-traditional agricultural export markets in the Caribbean. For Barrett, the evidence for staple food grains markets in sub-Saharan Africa was less abundant: 'The body of empirical evidence concerning smallholder staple food grains market participation patterns in eastern and southern Africa is thin but consistent and clear with respect to some basic descriptive patterns' (Barrett, 2008: 306). Generally, the literature on staples is weak and research into staple food crop marketing is undeveloped compared with that for the more exotic high-value markets. But for many smallholders, such domestic markets for staple food crops and livestock products are more important than export markets; they have the scale and linkages to poor households to permit broad-based agricultural and economic growth and reduce national poverty within a reasonable period of time.

Barrett found three dominant themes in the literature on Africa:

- a relatively small share of rural households sell staple food grains;
- there were strong associations between households' assets, especially of land, and geographic factors such as market access and agroecological zone – better-endowed households were much more likely to sell to market;
- weak institutional and physical infrastructure raised transaction costs and appeared to distort production and marketing behaviour.

Transaction costs

The role of institutions and transaction costs

A digression on institutions and transaction costs is important here. The introduction to economics of the study of transactions is attributed to the US political economist John R. Commons. In the tradition of the American Institutionalists analysing collective action, Commons was searching for an economic theory of the part played by collective action in the control of individual action. The three constituents of collective action were, he believed, conflict, dependence, and order. The unit of investigation that would encompass these three constituents was the transaction: 'so I made the transaction the ultimate unit of economic investigation, a unit of transfer of legal control' (Commons, 1934: 4).

Commons was one of the foremost members of what is now referred to as the Old Institutional Economics (OIE) school. OIE was a continuation of the 19th-century historical schools of political economy which mounted an attack on most aspects of the emerging neoclassical school and, above all, the behavioural assumptions associated with the notion of Rational Economic Man.

The fundamentals of OIE concern the organization and control of the economic system. The forces governing economic outcomes were regarded as mediated not first and foremost through the price mechanism, but through power relations, legal rights, and the role of the polity. The operation of the price mechanism was not disputed but institutions were held to supersede prices in importance: 'It is simply not true that scarce resources are allocated among alternative uses by the market ... The real determinant of whatever allocation occurs in any society is the organizational structure of that society – in short, its institutions' (Ayres, 1957: 26). Ayres, among others, was negative about the allocative role of institutions in economic activity, whereas Commons was more positive: institutions could promote economic development and well-being. In the aftermath of the Second World War, institutionalism gave way before the neoclassical renaissance which became the mainstream paradigm.

Nevertheless, the significance of institutions in influencing markets persisted, not least in Coase's influential 1937 and 1960 papers (Coase, 1937, 1960). Coase was primarily interested in the organization of the firm, but his comments on 'marketing costs' gave rise to the notion of transaction costs in addition to production costs of doing business. Arising out of this work, New Institutional Economics (NIE) – and particularly the Transaction Cost Economics (TCE) branch – was concerned with the organization and development of economic activity, of contractual arrangements between firms. Although Williamson traced the origins of TCE to the 1930s (Williamson, 1975), NIE derives its concern with transaction costs, and little else, from the OIE of Commons et al. Seminal contributions have been made to the development of NIE by many other writers.

Williamson's definition of the transaction serves as a first approximation: 'a transaction occurs when a good or service is transferred across a technologically separable interface. One stage of activity terminates and another begins' (Williamson, 1985: 1). Marion identifies four transaction elements (Marion and NC117 Committee, 1986):

- making the deal;
- transfer of ownership;
- establishing a price;
- physical delivery of the product to the buyer.

More generally a transaction is a process linking various functions, involving the exchange of information, goods, services, money, and property rights. Transaction costs are the costs of these exchanges. At the heart of transaction costs of 'measuring, monitoring and mediating' are the information and institutions that are the mechanisms to reduce uncertainty: efficient exchange means the need to 'measure' the attributes and characteristics of products and buyers and sellers, to 'monitor' agreements between buyers and sellers, mechanisms to 'mediate' in case of disputes between contracting parties. It is often said, citing Williamson and Coase, that TCE is an approach to the conceptualization and analysis of institutional, or contractual, arrangements – the diverse forms of contracting governing economic exchange – that are chosen in order to minimize transaction costs.

While drawing essential attention to the significance of non-price factors in determining decisions about buying and selling, this is fundamentally inaccurate. Farmers – and most other business people – do not act to minimize costs but to maximize revenue subject to costs (i.e. profit), including a risk threshold. Costs are both production costs and transaction costs, at the root of which is uncertainty. Profit is a function of product prices, quantities sold, and both production and transaction costs. Therefore, contractual arrangements are chosen (if there is choice) subject to the welfare outcomes of the costs and volumes of production, prices negotiated (if there is negotiation), and the transaction costs of alternative arrangements (if there are alternatives). Thus, in choosing contractual arrangements to maximize profit subject to idiosyncratic risk management preferences, for smallholder farmers 'contract design is a multi-criteria decision problem involving trade-offs' (Abebe et al., 2013: 15).

Farmers' costs

We have much more to learn about transaction costs. For example, Zanello et al. (2014) analysed transaction costs in farmers' market decision making in Ghana using a distinction between different types of transaction costs introduced by Key et al. (2000) and Bellemare and Barrett (2006):

> Proportional transaction costs are transaction costs that vary with the quantity traded. Often they are associated with the unit transport costs or the time required to make a sale. Fixed transaction costs are independent of the quantity traded and include the costs of seeking information on prices, costs of setting up a sale transaction and monitoring costs (that is, costs to ensure that the conditions of an exchange are met, for example enforcing the payment schedule) (Zanello et al., 2014: 1227).

Their results of the analysis of marketing behaviour in Ghana were at variance with results of other studies, leading them to conclude that factors driving farmers' decisions are likely to be significantly influenced by local institutional settings in addition to types of crops, size of transactions, and proximity to markets.

There is also potential for confusion when Barrett comments, 'The primary theme in the literature on smallholder market participation is the importance of transactions costs' (Barrett, 2008: 310). Writers confuse the physical costs of access to market with transaction costs. Admittedly, Coase (1937) did not use the term 'transaction costs', but 'marketing costs' – but it is unhelpful when Barrett writes that 'The transactions costs that have attracted most attention by analysts are those associated with transport' (Barrett, 2008: 310). Ownership, or lack, of a means of transport to market – donkey cart, bicycle, motorbike, pickup truck – is an important determinant of the ease of doing business, but is conceptually distinct from managing the uncertainty that is the root of transaction costs. In North's terms, transport costs are 'transformation' costs – or more commonly 'production costs' – rather than 'transaction' costs, and must be addressed by quite different sorts of intervention and initiatives.

Overall, reducing transaction costs is a key focus for policy intervention. In addition, interventions have been aimed at:

Farm level:

- increasing input supplies of fertilizer and seeds;
- stimulating the provision of credit;
- reducing the costs of physical access to local markets;
- organizing farmers.

Market level:

- policies to stimulate increased trader competition;
- integrating local and international markets.

As we come to consider market participation from a livelihoods perspective, it will be seen that these approaches are not exactly people-centred. Barrett himself draws attention to a number of key factors, consistent with the analyses in Chapter 1. He poses the question of whether barriers to market participation are primarily the lack of individual smallholders' privately held assets or the institutional and physical infrastructure of the external environment: 'This is an exceedingly important question that merits more attention from researchers' (Barrett, 2008: 314). However, it is one that is unlikely to generate a single, simple answer.

Households and livelihoods

Sen's entitlements

Besides economic modelling, Sen's entitlement approach (Sen, 1981) provides a way of considering rights to, and responsibilities for, natural-resources management and, particularly, economic supplies to households:

> Ownership relations are one kind of entitlement relations ... An entitlement relation applied to ownership connects one set of ownerships to another through certain rules of legitimacy. It is a recursive relation and the process of connecting can be repeated ...

Entitlement relations accepted in a private ownership market economy typically include the following, among others:

1. *trade-based entitlement:* one is entitled to own what one obtains by trading something one owns with a willing party (or, multilaterally, with a willing set of parties);
2. *production-based entitlement:* one is entitled to own what one gets by arranging production using one's own resources, or resources hired from willing parties meeting the agreed conditions of trade;
3. *own-labour entitlement:* one is entitled to own one's own labour power, and thus to the trade-based and production-based entitlements related to one's labour power;
4. *inheritance* and transfer entitlement: one is entitled to own what is willingly given to one by another who legitimately owns it, possibly to take effect after the latter's death (if so specified by him).

These are some entitlement relations of more or less straightforward kind, but there are others, frequently a good deal more complex. For example, one may be entitled to enjoy the fruits of some property without being able to trade it for anything else ... (Sen, 1981: 1–2).

Out of Sen's work, and in many subsequent contributions, much has been written about what has come to be known as the livelihoods approach to poverty reduction. Essentially, the assets-based approach conceives five classes of resources held at the individual, household, or collective levels, depending on the type of resource, on which there is general agreement among academics and practitioners (Donovan and Stoian, 2012). Nevertheless, there is no single understanding of assets, and, in particular, economists and anthropologists conceive them differently. What makes a livelihoods approach distinctive is the centrality of people, the human relations to productive assets through ownership and investment, and much-needed multidisciplinary perspectives.

For smallholder farmers, the following asset types are likely to be important, with particular emphasis on the social and human skills which have not been given sufficient weight in traditional analyses:

Livelihoods assets

Natural assets:

- land, water, livestock;
- production technologies such as new varieties;
- investments in resource conservation and management.

Human assets:

- technical skills for production;
- business and managerial skills.

Social assets:

- networks and access to markets and information;
- participation in collective activities associated with the economic activities of production, processing, marketing;
- participation in collective activities associated with democracy and governance;
- leverage from linkages with economic intermediaries.

Physical assets:

- domestic and productive buildings and equipment;
- tools and machinery.

Financial assets:

- access to credit;
- income benefit from product sales;
- access to longer-term investments.

Although livelihoods studies are fundamentally socioeconomic, there is benefit in re-envisioning the asset concepts from other perspectives, as Guyer (1997) makes clear. In an analysis of rural education among indigenous minorities in the Mexican state of Chiapas, a concept of cultural capital had particular significance (Bourdieu, 1983; Bebbington, 1999). Attempts have been made to reshape and expand the concept of social capital (Poole et al., 2013a), drawing attention to the interaction between cultural assets and education and communication technologies, thus depicting the key concepts for research not as a pentagon typically depicting five types of asset, but as a hexagon that also includes cultural capital (Figure 2.1).

This amplified conception of social and cultural capital in the livelihoods pentagon of assets is affirmed by IFAD in its *Rural Development Report 2016* (IFAD, 2016): the report casts a spotlight on indigenous peoples and acknowledges their unique characteristics and vulnerabilities:

> Their traditional knowledge, holistic practices and production systems both provide for sustainable management of resources and ensure that biodiversity is maintained for future generations. Recognizing how indigenous peoples have been able to make social capital, agriculture and the environment work together over centuries is crucial to an understanding of inclusive rural transformation, and offers an opportunity to learn from their sustainable livelihood practices (IFAD, 2016: 338).

Other livelihoods-capital considerations

Besides ownership of and investment in assets, the relationship of people to assets and the use of assets by people can be marked by other phenomena,

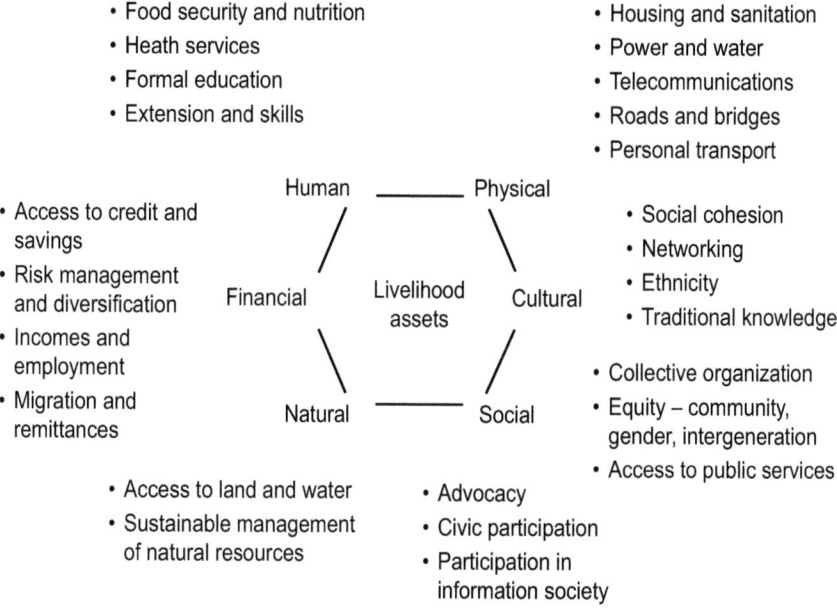

Figure 2.1 Livelihoods assets hexagon framework for agricultural development
Source: adapted from Poole et al. (2013a)

less well substantiated in the livelihoods literature, that are closely linked to the dynamics of household management:

Investment. The 'anthropology of wealth' is a counterpart to the economics focus on assets and poverty (Guyer, 1997). Guyer points out the gap between an anthropologist's concern with dynamic resource management and allocation, essentially accumulation, and the static nature of economic models which 'do not show people investing in [assets]: maintaining, increasing, scheming, and planning' (Guyer, 1997: 119). Essentially, Guyer wants consideration of (ongoing strategies of) investments as well as (starting) endowments, differentiating within the household between men and women. Notwithstanding, changes in assets have been explored more recently from a socioeconomic perspective, at household and institutional level (Donovan and Stoian, 2012; Donovan and Poole, 2013, 2014, 2016; Poole and Donovan 2014). Investments in social relationships, which are admittedly difficult to quantify, have insurance value as well as making an intrinsic contribution to well-being. They are polysemic (have multiple meanings), as are children who are consumption and investment assets, depending on generational perspectives. Similarly, investments in 'health' (through improved nutrition and disease prevention), directly (through food and medicines) and indirectly (through improved housing and sanitation), are at the same time consumption assets, investment assets, and reduce vulnerability to shocks.

Interactions and trade-offs. Exercising choice between making investments in new activities and pursuing only traditional activities involves potential costs.

At the simplest level, it could be the switch from subsistence food production to commercial production. What is the opportunity cost of the project activity? How has the relationship between subsistence/food prices and prices of the new commercial products evolved? The demand for labour is an important element in the shift towards more intensive production of higher-value products. Similarly, the increased cost of inputs and technology for higher-value production will affect borrowing requirements, cash flows, and other household spending decisions. There is a generational perspective to the trade-offs: for example, children can be asked or compelled to work in order to bring in immediate financial assets, or sent to school to enhance long-term social capital.

Convertibility. Guyer (1997) also points out that assets are 'multifaceted'; that is, they can be put to more than one use. In addition, assets are also convertible, financial assets being the most fungible: quickly convertible, for example, into physical capital (equipment and housing), natural capital (land and livestock), social capital (gifts and prestige goods), and more slowly convertible into human capital through education. But financial assets are also the most vulnerable to rapid depletion (loss of cash, loss of access to credit).

Asset erosion. It is possible that small producers are left impoverished from shocks. Vulnerability and risk are critical factors which can lead to asset erosion. Apart from household affairs like illness, accidents, and death, there are natural disasters like floods and drought that can impact severely on the household economy. It maybe through individual misjudgements or unwise development initiatives that farmers take credit that they cannot repay; maybe the market collapses and the investments made in new opportunities become sunk costs, i.e. costs that cannot be recovered. Investment by farmers in crop insurance is a mechanism for avoiding massive losses by making relatively small expenditures that in the best circumstances will not be recovered – at least financially – but will reduce household vulnerability. Specific culture in the ethnographic sense can be conceived of as an asset that confers stability and coherence, as for example in ethnic-minority communities (Poole et al., 2013a). Cultural change can erode such cultural capital. But cultural change can also build other assets: a decline in traditional mores may erode social stability and investment in female seclusion and consequent investment in marriage, but also offer new forms of investment, for example through girls' education and female employment – even engagement in agribusiness marketing.

Sustainability. Assessing natural capital is important for considering environmental and other impacts and the continuity of an enterprise; an attempt at environmental assessment can be made through investment in indicators of natural capital (e.g. soil, water, and waste management), but this is a *very* complex and specialist area of analysis. It is probably though qualitative data collection that the long-term perspective of changes can be best assessed. There is also a human, social, and economic dimension to sustainability.

Whereas interventions promoting new technology might achieve adoption and sustainability within the short-medium term, say two to four years, sustainability of initiatives to do with specific human skills and economic organizations such as cooperatives are likely to take longer, say five years or more, and therefore is not likely to be observable. Economic or business sustainability will be significantly dependent on the external environment over which smallholders have no control.

Targeting. There is a question about which smallholder households actually benefit from external interventions. Furthermore, who is excluded? Are there entry or participation thresholds in the project design? Does a given project only help those smallholders who are better off and easier to help, or was the capacity of the really poor smallholders to participate in markets enhanced, irrespective of age, gender, education, and scale?

Asset-based approaches are now commonly linked to value chain approaches, and many analytical tools blend the basic concepts. Guyer can be cited again to orient the empirical work this book reports:

> Since the sociocultural anthropology of the last 25 years is an anthropology that has been increasingly focused on dynamics and history (valuation rather than values, cultural construction rather than the structure of culture), the study of assets would become not only the study of the assets themselves, or of asset management (in the life-cyclical sense) but of asset creation (in the active, historical sense). Both policies and popular processes create assets ... (Guyer, 1997: 123).

Value chain thinking

Development of the chain concept

Value chain approaches have become the norm for market analyses and market projects and interventions. The empirical work reported in this book was conducted within a value chain framework (Morrison, 2010), and such concepts are used in the analysis and conclusions. In explaining value chain thinking, we note that there is a burgeoning literature on agri-food value chains in developing and emerging markets that has matched the explosion of interest in the value chain concept and value chain approaches to development and poverty reduction in poorer countries since 2000. However, not all things to do with value chains are new. The discussion that follows draws on selected literature including a review published in the journal *Food Chain* (Poole, 2013).

Application of the chain concept to agri-food systems has a number of roots. As noted above, Coase's seminal work in the 1930s (1937) that gave rise to transaction-cost economics was, in fact, concerned with and used the term marketing costs. He recognized that markets were coordinated not only by the price mechanism in spot markets but also by strategic decisions of managers in firms to either 'make or buy' in terms of sourcing and distribution of products and services. Under certain circumstances, efficient inter-firm linkages were preferred over the integration of economic functions within a single firm, and linkages – or multiple links within a chain of transactions – became a focus of analysis. Thus it became evident that marketing chains should be considered as a common phenomenon in industrial organization.

Similarly, as noted above, agricultural economists have long analysed the efficiency with which products have reached markets in terms of prices, margins, and costs. Concern beyond the price mechanism about vertical coordination between agribusiness market actors was evident in the United States in the 1960s. Concerns about market efficiency were succeeded by a growing interest in the institutional environment and inter-firm contractual arrangements through the lens of New Institutional Economics (Williamson, 1985; Coase, 1988; North, 1990).

The emergence from industrial organization of value chain analysis as an analytical framework can be traced, inter alia, to Porter's exposition of competitive strategies (1985), which provided a tool to enable firms to look beyond their own boundaries, to examine linkages with other organizations, and to identify ways of creating and sustaining better business performance. Porter's contribution was to recognize the importance of the processes of adding value to the flow of goods and services from suppliers through intermediaries to final consumers in enhancing firm profitability. His focus on creating and sustaining firm and national competitive advantages transformed the economics of industrial organization into a more business-friendly explanation of firm and industrial strategy (Porter, 1985, 1990). The marriage of chain perspectives on inter-firm organization with the insistence on value addition for competitive advantage gave birth to the value chain concept (Porter, 1985; Kaplinsky, 2000).

From a different school of thought, world-systems theory gave rise to chain-type thinking in terms of international trade and global commodity/value chains, focusing on the loci of power in chain management being either buyer-led or supplier-led (Gereffi and Korzeniewicz, 1994; Gibbon, 2001, 2003). The French *filière* school (Griffon, 1989; Tallec and Bockel, 2005) developed a parallel tradition focusing on the efficiency of operations, flows of resources, and interdependencies within vertical commodity chains. Each framework had something to offer and each something to learn (Raikes et al., 2000).

As a result of the convergence of these threads, in the last decade, value chains have become the dominant discourse of many governments, international donors, NGOs, and research on rural markets in developing economies (Humphrey, 2005). It is not possible or necessary to distil all the literature into a single review, but a synthesis of the key concepts is presented in Figure 2.2.

Terminology matters. *Value chain approaches* are interventions by public sector and non-governmental organizations and initiatives by the private

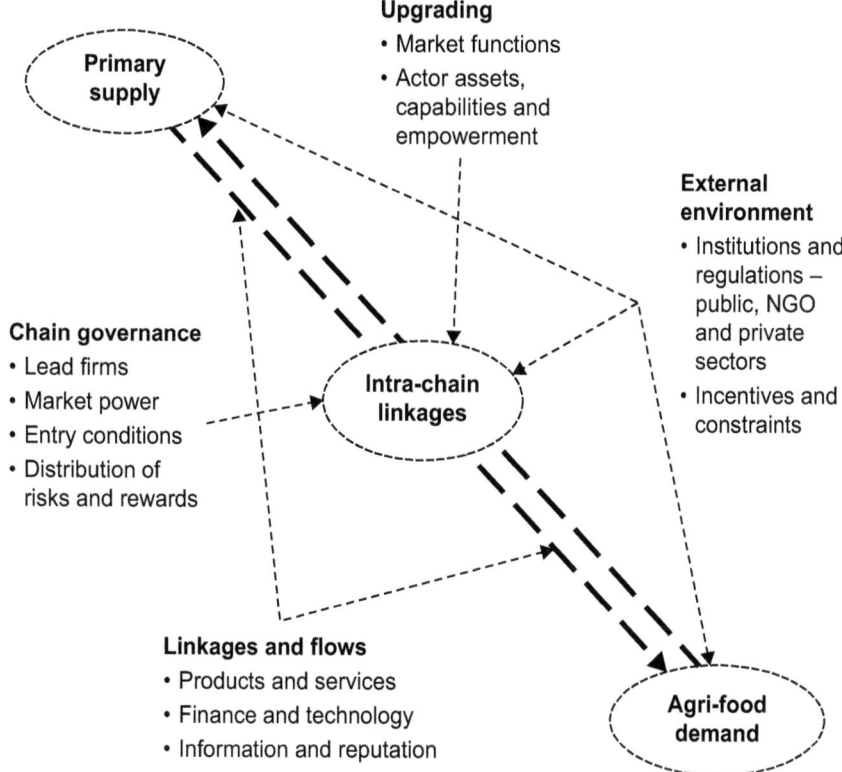

Figure 2.2 Concepts underpinning the agri-food value chain
Source: adapted from Poole (2013: 200)

sector, to overcome chain constraints, align incentives, and promote development. A definition which will satisfy most analysts is:

> a *value chain* comprises the linkages between actors and the flows of products, services, resources and information among economic actors: households, firms and other organizations such as cooperatives

Key features of *value chain research* are analyses of:

- governance – the formal and informal relationships (vertical and horizontal) between chain actors, and the exercise of power, which facilitate entry and shape risks and rewards;
- upgrading – value addition resulting from interventions and initiatives to improve chain functions and actors' capabilities and empowerment;
- the distributional outcomes, primarily economic;
- the influence of the external environment of politics, socioeconomics, culture, and technology, and the role of salient public, private, and third-sector organizations, and of formal laws, regulations, and standards and informal norms of behaviour.

Value chain interventions are development activities targeted at specific points in the chain, or even the whole chain, from production to consumption, that aim to overcome imperfections in efficiency, equity, or some other dimension.

Setting 'systems' boundaries for value chain research has been necessary to limit analysis to tractable questions and feasible empirics, and, consequently, analyses have commonly been case studies, centred on business dyads rather than whole chains. Thus, whole chain analyses have been few. The study reported in IFPRI (2010) may be the acme of value chain analyses, but employed a level of resources that would be unimaginable to most researchers.

Recent literature on value chains for development

There is no single approach to value chains. Following Porter, the value chain has been used as a conceptual framework for analysing participation, governance, equity, and the outcomes of business activities thus organized. NGOs have become a significant channel for linking poor people to markets and, together with international donor-practitioners, have drawn on the value chain approaches more commonly used in advanced economies and international commodity exports. Thus, various tools have been developed to help improve smallholder participation in markets, and these are targeted primarily at projects and initiatives attempting to improve the situation of smallholder producers and small enterprises in the value chain.

Beyond poverty reduction, the scope for value chain research extends to a range of policy issues in economic growth and agricultural development. Consideration of some of these is, at best, in its infancy: research and input supply industries serving primary production; domestic staple foods as well as exotic, niche exports; hunger, health and food safety; gender; climate and environmental sustainability; carbon flows; water management; scalability of interventions; labour and social responsibility; the public sector as an actor within and without the chain; and state-building.

No feasible research can tackle all these issues, but still there are gaps:

- Curiously, consumers are often not considered to be chain actors (Hawkes, 2009). Where food consumption is an end in itself, as well as a means to other ends such as employment creation and economic growth, this is inexplicable.
- Employment per se, as an important multiplier in economic development in general and the agri-food sector in particular, has been strangely neglected. For poverty reduction in rural areas where landless labourers are often the most disadvantaged, like much of South Asia, this is even more surprising.
- Actually, food ends as nutrition, and there is an upsurge of global interest in the impact of agri-food chains on nutrition and health

outcomes – for the poor and the rich, the underfed and overfed, developing and advanced economies, and cross-cutting contexts where all these dimensions may coexist.

Recognizing that conflict is one of the major factors creating and sustaining global poverty and inequality, building viable economic activity in post-conflict situations has to be a focus for new research.

Analysis of the sustainability dimensions of value chains needs to draw on the techniques of environmental impact life-cycle and analyses, with carbon management being a priority.

Beyond pollution, agri-food waste and disease management can also be added to the agenda for value chain analysis.

Chains are pathways through more general and enduring market systems. They may come and go but rarely stay the same, and dynamic analyses are also necessary, especially in fast-evolving consumer markets. Much research – and value chain analyses are no exception – adopts a static case-study approach. Research needs to capture the dynamic nature of markets: not just intervention impacts over months, but changing supply and demand conditions over years.

Returning to themes that are central to this book, on upgrading and inclusion, three initial considerations must be made explicit: firstly, is participation in commercial markets always a desirable objective? Secondly, can upgrading have the perverse effect of raising entry barriers and excluding the poor whose improvement is being targeted? Questions of equality and gender underlie this preoccupation. And thirdly, note that a given chain may be less important to the actors themselves than to the keen researchers; people's livelihoods are diverse, and thus a given chain may be only one in a portfolio of activities. Livelihood context matters.

This book follows two recent value chain publications and, it is hoped, adds helpfully to the growing literature rather than duplicates existing efforts. In their recent volume, Coles and Mitchell (2011: 15) set out an immodest agenda: 'This book seeks to address one of the most intractable contemporary development challenges – what can the billion poorest people do to improve their livelihoods and join the trend of rising prosperity in the developing world?' The answer given is, upgrading their position in a range of natural resource-based value chains.

The other book is a synthesis, edited by the Overseas Development Institute, London, of research that has emerged from a programme of the International Development Research Centre, Canada, that aimed to integrate poverty, gender, and environmental concerns into value chain research and increase incomes for the rural poor in a sustainable manner. The research was undertaken between 2007 and 2009, in South and East Asia (India, Nepal, Vietnam, and the Philippines) and sub-Saharan Africa (Mali, Tanzania, and Senegal), drawing on the experiences of farmers, development workers, and policymakers 'from the South'.

Upgrading

Different types of upgrading are commonly recognized: horizontal and vertical coordination; 'doing different things' – functional upgrading; product and process upgrading; and skills transfer – inter-chain upgrading. A range of lessons is derived concerning, inter alia:

- the inappropriate imposition of structures from outside;
- the existence of trade-offs between objectives;
- the complexity of balancing costs and benefits of novel contracting;
- whether responsive trading relationships which reduce vulnerability and uncertainty – non-financial benefits – can coexist alongside imbalances in market power;
- specialization to reduce the costs and risks of functional integration;
- the advantages and disadvantages of working closely with intermediary firms;
- the role of state intervention to overcome market failures due to ineffective standards, lack of rural infrastructure, and costly logistics;
- costs and benefits of certification;
- the significance of price-quality disconnects for poverty and environmental impacts;
- the complexity of labour issues;
- the advantages of inter-chain upgrading arising from economies of scope.

The concluding chapter claims that 'our approach has been intensely practical ... this book examines research-based evidence for the effectiveness of upgrading interventions' (Coles et al., 2011: pp. 236–7). It adds usefully to the theoretical literature on value chain analysis, but less on experiences of value chain interventions. Perhaps the cases are little more than typical market development interventions. One might have expected more on the critical role of human capacity building and on empowerment, which is mentioned in a couple of places but tangentially, or apparently as an afterthought on gender, but not as a fundamental element of the framework.

Participation: principles and determinants

There are three principal themes to the volume edited by Helmsing and Vellema (2011). Firstly, on governance and inclusion: including the poorest in poverty reduction interventions must be tested against the trade-offs and risk. As is argued later, 'inclusion and exclusion are more usefully seen as processes shaping *how* (rather than *if*) actors participate' (Helmsing and Vellema, 2011: 12). Some actors may be excluded or 'self-exclude' from a particular chain in favour of other activities and networks. Thus, targeting 'beneficiaries', and the scope for upscaling successful interventions, become important policy dilemmas, and voice should be given to the 'targets' to consider strategic choices – such as whether or not to adopt production certification and product standards. Moreover, contextual factors such as public policy and industry organization are likely to mediate different welfare outcomes from the same choice sets among value chain-driven intervention mechanisms. Cooperatives are a favoured development approach, but the benefits of formal democratic governance become attenuated by community mechanisms and hierarchical arrangements as the need for scale and efficiency increases.

Including labour-market institutions within the value chain encompasses stakeholders who are less visible than producer-entrepreneurs and downstream firms. Yet employees may be among the poorest of the poor, and recent research shows that the benefits of fair trade do not extend into the labour market, at least in Ethiopia and Uganda (Cramer et al., 2014). Labour-process theory integrates into value chain analysis the relations of bargaining power, seasonality of rural activities and labour demand, the formalization and organization of contracts, and the establishment of pay rates and returns. One would expect gender relations to be similarly incorporated, but this is not done here. The importance of including the state as a principal actor in the chain, as well as a proximate player in the external environment, is highlighted by the case of the [semi-] liberalized Ghanaian cocoa sector. There, the state is an intermediary which attenuates somewhat the market signals for social and environmental issues.

Secondly, the emphases on embedding and business systems concern behaviour as a function of the socio-cultural and political context, and the interrelationships between, and contrasting incentives among, chain actors and non-chain actors. Thus, embeddedness can be observed at the level of the network or chain, the society, or some wider (geographic) notion of 'territory'. The contribution of business-systems theory is to recognize that even global chains are embedded within specificities of territorial institutions, culture, and industry. Within this view, a strong role is envisaged for the state in defining pro-developmental institutions (rather than, say, the private sector as a facilitating agent), including the regulation and management of natural resources and government support in case of shocks.

Power, politics, and profit distribution are features of business systems that may or may not be aligned with incentives to participate or exclude individual actors. They may entrench the economic positions of elites rather than promote development by permitting or encouraging innovation and competition. Alternatively, lead firms may condition business relationships in a manner consistent with environmental and social responsibility.

Finally, for chain-based partnerships for development, the premise is that 'access to assets is a necessary but not sufficient condition for escaping poverty' (Helmsing and Vellema, 2011: 15). First, partnerships are necessary and are viewed as having horizontal and vertical dimensions, both within the chain and also with actors from the wider business environment. New institutional relationships can help poor actors overcome the constraints common to smallholder participation, such as lack of information, lack of market alternatives, lack of finance, and small scale. Donors and NGOs may intervene through narrowly defined projects at specific points in the value chain but often ignore these wider partnership constraints. Second, having achieved some measure of improved participation, the authors note that 'the literature pays surprisingly little attention to issues of upscaling' (Helmsing and Vellema, 2011: 16).

Developing an integrative framework, rather than using a singular (economic) conception of what a value chain is, allows diverse researchers to reveal the different meanings and incentives found in local communities. Aligning the incentives arising from alternative understandings is necessary to formulate favourable conditions of participation for vulnerable groups – and hence poverty reduction.

> This volume argues that the development impacts of inclusion of small producers, local firms and workers in (global) value chains importantly depends on two conditions: the terms of participation in the process of inclusion and the degree of alignment of value chain logics with the capacities of actors and the institutions embedded in territorial business systems (Helmsing and Vellema, 2011: 18).

Much literature on value chains has not moved beyond case studies. In its conceptualization, this book does. The multidisciplinary approach has yielded important insights (something that this author has signalled elsewhere; Poole et al., 2013b). Measured against the framework introduced above, there is much work still to be done, and that is as it should be.

References

Abebe, G.K., Bijman, J., Kemp, R., Omta, O. and Tsegaye, A. (2013). Contract farming configuration: smallholders' preferences for contract design attributes. *Food Policy* **40**: 14–24 <http://doi.org/10.1016/j.foodpol.2013.01.002>.

Ayres, C.E. (1957). Institutional economics: discussion. *American Economic Review* **47**: 26–27.

Barrett, C.B. (2008). Smallholder market participation: concepts and evidence from eastern and southern Africa. *Food Policy* **33**(4): 299–317 <http://doi.org/10.1016/j.foodpol.2007.10.005>.

Bayudan-Dacuycuy, C. (2013). The influence of living with parents on women's decision-making participation in the household: evidence from the Southern Philippines. *Journal of Development Studies* **49**(5): 641–656 <http://dx.doi.org/10.1080/00220388.2012.682987>.

Bebbington, A. (1999). Capitals and capabilities: a framework for analyzing peasant viability, rural livelihoods and poverty. *World Development* **27**(12): 2021–2044 <https://doi.org/10.1016/S0305-750X(99)00104-7>.

Becker, G.S. (1976). Altruism, egoism, and genetic fitness: economics and sociobiology. *Journal of Economic Literature* **14**(3): 817–826.

Bellemare, M. and Barrett, C.B. (2006). An ordered tobit model of market participation: evidence from Kenya and Ethiopia. *American Journal of Agricultural Economics* **88**: 324–337 <http://doi.org/10.1111/j.1467-8276.2006.00861.x>.

Bennett, F. (2013). Researching within-household distribution: overview, developments, debates, and methodological challenges. *Journal of Marriage and Family* **75**(3): 582–597 <http://dx.doi.org/10.1111/jomf.12020>.

Bourdieu, P. (1983). 'Forms of capital', in J.C. Richards, ed., *Handbook of Theory and Research for the Sociology of Education*. New York, Greenwood Press.

Coase, R.H. (1937). The nature of the firm. *Economica* **4**(November): 386–405.

Coase, R.H. (1960). The problem of social cost. *Journal of Law and Economics* **3**(October): 1–44.

Coase, R.H. (1988). *The Firm, the Market and the Law*. Chicago, University of Chicago Press.

Coles, C. and Mitchell, J., eds (2011). *Markets and Rural Poverty: Upgrading in value chains*. London and Ottawa, Earthscan and International Development Research Centre.

Coles, C., Mitchell, J., Owaygen, M. and Shepherd, A. (2011). Conclusions, in C. Coles and J. Mitchell, eds, *Markets and Rural Poverty: Upgrading in value chains*, pp. 235–260. London and Ottawa, Earthscan and International Development Research Centre.

Commons, J.R. (1934). *Institutional Economics: Its place in political economy*. New York, Macmillan.

Cramer, C., Johnson, D., Oya, C. and Sender, J. (2014). *Fairtrade, Employment and Poverty Reduction in Ethiopia and Uganda*. Final Report to DFID, April. London, SOAS, University of London.

Dauphin, A., El Lahga, A.-R., Fortin, B. and Lacroix, G. (2011). Are children decision-makers within the household? *The Economic Journal* **121**(553): 871–903 <http://dx.doi.org/10.1111/j.1468-0297.2010.02404.x>.

DFID (2013). *Market Systems Development Platform - April 2013. Annex A : Extract from the Market Systems Development Platform Business Case. Intervention Summary*. London, Department for International Development (DFID).

Donovan, J. and Poole, N.D. (2008). *Linking Smallholders to Markets for Non-traditional Agricultural Exports: A review of experiences in the Caribbean Basin. Inputs for strategy formulation by 'EU-ACP All Agricultural Commodities Programme'*. EU-AAACP Paper Series. No. 2. Rome, FAO. Retrieved 28 March 2017, from http://www.fao.org/fileadmin/templates/est/AAACP/caribbean/FAO_AAACP_Paper_Series_No_2_1_.pdf.

Donovan, J. and Poole, N.D. (2013). Asset building in response to value chain development: lessons from taro producers in Nicaragua. *International Journal of Agricultural Sustainability* **11**(1): 23–37 <http://dx.doi.org/10.1080/14735903.2012.673076>.

Donovan, J. and Poole, N.D. (2014). 'Changing asset endowments and smallholder participation in higher value markets: evidence from certified coffee producers in Nicaragua'. *Food Policy* **44**: 1–13 <https://doi.org/10.1016/j.foodpol.2013.09.010>.

Donovan, J. and Poole, N.D. (2016). 'Changing asset endowments and smallholder participation in higher-value markets: evidence from certified-coffee producers in Nicaragua', in A. Devaux, M. Torero, J. Donovan and D. Horton, eds, *Innovation for Inclusive Value Chain Development: Successes and challenges*, pp. 93–126. Washington, DC, International Food Policy Research Institute (IFPRI).

Donovan, J. and Stoian, D. (2012). *5 Capitals: A tool for assessing the poverty impacts of value chain development*. Technical Series 55, Rural Enterprise

Development Collection no. 7. Turrialba, Costa Rica, CATIE. Retrieved 28 March 2017, from http://www.worldagroforestry.org/publication/5-capitals-tool-assessing-poverty-impacts-value-chain-development.

Dorward, A.R. and Poole, N.D. (2004). *Functioning of Markets and the Livelihoods of the Poor*. Making Markets Work Better for the Poor. Proceedings of the Inception Workshop, November 2003. Hanoi, Asian Development Bank.

Doss, C. (2013). Intrahousehold bargaining and resource allocation in developing countries. *World Bank Research Observer* **28**(1): 52–78.

Gereffi, G. and Korzeniewicz, M., eds (1994). *Commodity Chains and Global Capitalism*. Westport, CT, Greenwood Press.

Gibbon, P. (2001). Upgrading primary production: a global commodity chain approach. *World Development* **29**(2): 345–363 <https://doi.org/10.1016/S0305-750X(00)00093-0>.

Gibbon, P. (2003). The African Growth and Opportunity Act and the global commodity chain for clothing. *World Development* **31**(1): 1809–1827 <https://doi.org/10.1016/j.worlddev.2003.06.002>.

Griffon, M., ed. (1989). *Economie des filières en régions chaudes. Formation des prix et échanges agricoles. Actes du Xeme séminaire d'économie et de sociologie, 11–15 septembre*. Montpellier, France, CIRAD.

Grossbard, S. (2010). *Independent Individual Decision-Makers in Household Models and the New Home Economics*. IZA Discussion Paper No. 5138. Bonn, Germany.

Guyer, J.L. (1997). 'Endowments and assets: the anthropology of wealth and the economics of intrahousehold allocation', in L. Haddad, J. Hoddinott and H. Alderman, eds, *Intrahousehold Resource Allocation in Developing Countries: Models, methods, and policy*, pp. 112–125. Baltimore and London, Johns Hopkins University Press.

Haddad, L., Hoddinott, J. and Alderman, H., eds (1997). *Intrahousehold Resource Allocation in Developing Countries: Models, methods, and policy*. Baltimore and London, Johns Hopkins University Press.

Hall, P.A. and Soskice, D. (2001). *Varieties of Capitalism: the institutional foundations of comparative advantage*. Oxford, OUP.

Harper, M., Belt, J. and Roy, R., eds (2015). *Commercial and Inclusive Value Chains*. Rugby, UK, Practical Action Publishing.

Hawkes, C. (2009). Identifying innovative interventions to promote healthy eating using consumption-oriented food supply chain analysis. *Journal of Hunger & Environmental Nutrition* **4**(3/4): 336–356 <http://dx.doi.org/10.1080/19320240903321243>.

Helmsing, A.H.J. and Vellema, S., eds (2011). *Value Chains, Social Inclusion and Economic Development: Contrasting theories and realities*. London, Routledge.

Humphrey, J. (2005). *Shaping Value Chains for Development: Global value chains in agribusiness*. Eschborn, Germany, GTZ.

IFAD (2016). *Rural Development Report 2016: Fostering inclusive rural Transformation*. Rome, United Nations International Fund for Agricultural Development. Retrieved 28 March 2017, from https://www.ifad.org/ruraldevelopmentreport.

IFPRI (2010). *Pulses Value Chain in Ethiopia: Constraints and opportunities for enhancing exports*. Working Paper. Washington DC, International Food Policy Research Institute (IFPRI).

Jones, W.O. (1974). Regional analysis and agricultural marketing research in tropical Africa: concepts and experiences. *Food Research Institute Studies* **13**(1): 3–28.

Kaplinsky, R. (2000). *Spreading the gains from globalisation: what can be learned from value chain analysis. IDS Working Paper No. 10.* Brighton, Institute of Development Studies.

Key, N., Sadoulet, E. and de Janvry, A. (2000). Transactions costs and agricultural household supply response. *American Journal of Agricultural Economics* **82**: 245–259 <https://doi.org/10.1111/0002-9092.00022>.

Li, L. and Wu, X. (2011). Gender of children, bargaining power, and intra-household resource allocation in China. *Journal of Human Resources* **46**(2): 295–316 <http://dx.doi.org/10.3368/jhr.46.2.295>.

Lundberg, S., Romich, J.L. and Tsang, K.P. (2009). Decision-making by children. *Review of Economics of the Household* **7**(1): 1–30 <http://dx.doi.org/10.1007/s11150-008-9045-2>.

Marion, B.W. and NC117 Committee (1986). *The Organization and Performance of the US Food System.* Lexington, MA, Lexington Books.

Morrison, J.A. (2010). *Brief guide prepared in support of capacity development workshops on the implementation and use of value chain studies approaches for policy analysis.* Rome, UN Food and Agriculture Organization.

North, D.C. (1990). *Institutions, Institutional Change and Economic Performance.* Cambridge, Cambridge University Press.

Poole, N.D. (2010). 'From 'marketing systems' to 'value chains': what have we learnt since the post-colonial era and where do we go?' In H. van Trijp and P. Ingenbleek, eds, *Markets, Marketing and Developing Countries: Where we stand and where we are heading*, pp 18–22. Wageningen, the Netherlands, Wageningen Academic Publishers.

Poole, N.D. (2013). Value chain perspectives and literature: a review. *Food Chain* **3**(3): 199–211 <https://doi.org/10.3362/2046-1887.2013.019>.

Poole, N.D. and de Frece, A. (2010). *A Review of Existing Organisational Forms of Smallholder Farmers' Associations and their Contractual Relationships with other Market Participants in the East and Southern African ACP Region.* EU-AAACP Paper Series. No. 11. Rome, Food and Agriculture Organization of the United Nations. Retrieved 28 March 2017, from http://www.fao.org/fileadmin/templates/est/AAACP/eastafrica/FAO_AAACP_Paper_Series_No_11_1_.pdf.

Poole, N.D. and Donovan, J. (2014). Building cooperative capacity: the speciality coffee sector in Nicaragua. *Journal of Agribusiness in Developing and Emerging Economies* **4**(2): 133–156 <http://dx.doi.org/10.1108/JADEE-01-2013-0002>.

Poole, N.D., Alvarez, F., Vazquez, R. and Penagos, N. (2013a). Education for all and for what? Life-skills and livelihoods in rural communities. *Journal of Agribusiness in Developing and Emerging Economies* **3**(1): 64–78 <http://dx.doi.org/10.1108/20440831311321656>.

Poole, N.D., Chitundu, M. and Msoni, R. (2013b). Commercialisation: a meta-approach for agricultural development among smallholder farmers in Africa? *Food Policy* **41**(August): 155–165 <https://doi.org/10.1016/j.foodpol.2013.05.010>.

Porter, M.E. (1985). *Competitive Advantage: Creating and sustaining superior performance.* New York, Macmillan.

Porter, M.E. (1990). *The Competitive Advantage of Nations.* New York, Macmillan.

Raikes, P., Jensen, M.F. and Ponte, S. (2000). Global commodity chain analysis and the French filière approach: comparison and critique. *Economy and Society* **29**(3): 390–417 <http://dx.doi.org/10.1080/03085140050084589>.

Sen, A. (1981). *Poverty and Famines: An essay on entitlements and deprivation.* Oxford, Clarendon.

Tallec, F. and Bockel, L. (2005). *L'approche filière: analyse fonctionelle et identification de flux. EASYPol Présentation Thématique Générale Module 043.* Rome, Food and Agriculture Organization of the United Nations.

The Springfield Centre (2008). *Perspectives on the Making Markets Work for the Poor (M4P) Approach.* Berne, Swiss Agency for Development and Cooperation SDC. Retrieved 22 May 2017, from http://www.value-chains.org/dyn/bds/docs/681/Perspectives%202008.pdf.

The Springfield Centre (2015). *The Operational Guide for The Making Markets Work for the Poor (M4P) Approach.* London/Berne, Funded by Swiss Agency for Development and Cooperation and Department for International Development. Retrieved 22 May 2017, from http://www.enterprise-development.org/wp-content/uploads/m4pguide2015.pdf.

Tincani, L. (2012). Resilient Livelihoods: Adaptation, Food Security and Wild Foods in Rural Burkina Faso. Unpublished PhD. London, SOAS, University of London.

Williamson, O.E. (1975). *Markets and Hierarchies: Analysis and antitrust implications.* London, Macmillan.

Williamson, O.E. (1985). *The Economic Institutions of Capitalism.* New York, Macmillan.

World Bank (2016). *Doing Business 2017: Equal opportunity for all.* Washington, DC, World Bank Group. Retrieved 28 March 2017, from http://www.doingbusiness.org/.

World Bank Group (2016). *Enabling the Business of Agriculture 2016: Comparing Regulatory Good Practices.* Washington, DC, World Bank. Retrieved 28 March 2017, from http://eba.worldbank.org/~/media/WBG/AgriBusiness/Documents/Reports/2016/EBA16-Full-Report.pdf.

Zanello, G., Srinivasan, C.S. and Shankar, B. (2014). Transaction costs, information technologies, and the choice of marketplace among farmers in Northern Ghana. *Journal of Development Studies* **50**(9): 1226–1239.

CHAPTER 3
Financial services for agricultural smallholders

This chapter discusses finance, which is one of the biggest barriers for smallholder farmers wishing to engage in markets. On the one hand, farmers can easily fall into debt; on the other, it is difficult for farmers to access financial capital to expand production and marketing. The chapter reviews diverse and innovative forms of financing that can improve services in many circumstances.

The transaction costs introduced in Chapter 2 are found to be significant in the delivery of financial services. We deepen the discussion of the behavioural features of financial markets at the heart of which are informational problems: lack of good information increases financial risks. The chapter concludes with a discussion of agricultural development strategies that can link credit provision with the delivery of other forms of market development support in ways that reduce risks for lenders and stimulate agricultural production and marketing. A final caveat is that implanting sustainable models of integrated agricultural development services can take years.

Keywords: agricultural finance, credit, transaction costs, information, behaviour, risk

Background

The role of finance

Looking back to the early post-colonial era of the 1950s and 1960s, many international organizations and governments in developed and developing countries considered that the prevailing markets for agricultural finance were a constraint to boosting the productivity of the rural sector and to fostering the adoption by farmers of new technologies and inputs. Traditional, and often long-standing, systems of finance were considered to be exploitative and inefficient. The provision of credit would help to overcome the key constraints.

Views about market performance in those days owed more to preconceptions than evidence – at least on the output side (Southworth et al., 1979; Holtzman, 1989). Nevertheless, research did identify practices that damaged agricultural development. For example, Crow's study of the rice trade in Bangladesh uncovered complex market structures and interrelationships between growers and traders who were often moneylenders (Crow, 1989). Crow considered credit to be pivotal to understanding the operations of the rice market. Interest rates were largely implicit but very high – into the hundreds of per cent – and commissions had to be paid by farmers before credit could be obtained. The commission agents were identified as the foci of market power. Of particular interest to food policy analysts was the impotence of the government of a food-scarce country in attempting to influence producer returns and hence alter the incentives to increase production for the market and so promote national food security. The system of *dadon* – forward sales as a means of financing production by farmers – provided 'bonded' suppliers for the market participant, thereby securing rice supply sources at favourable terms, and assisted in creating personal and long-term market relationships. Crow found that millers also enjoyed a high degree of market power, to the extent that they could even hold up government rice policy until it changed in their favour; that is, until the government, in the interests of somehow getting supplies moving through the system to assist national food security, was forced to alter its buying policy in favour of the millers.

Officials were found to actually collude in the circumvention of government policy by diverting public rice supplies to private markets. According to Crow (1989: 217), 'any policy which tends to stabilize prices may be expected to reduce returns to a section of the trading community'. Hence it was in the millers' interests to undermine the success of the formal supply channel. The hidden vertical relationships concentrated market power among a merchant elite who definitely were not price takers, but who exerted considerable market power through interlocked markets – 'the power of merchants and landlords as the main suppliers of credit' (Crow, 1989; 222).

Such analyses, critical of often informal and potentially exploitative financial markets for smallholder production, raised many questions about how credit relations affected poor farmers and food-security objectives. Accordingly, many governments addressed market failures by intervening in credit provision by limiting interest rates, allocating lending quotas for poor farmers to formal finance institutions, launching and providing capital to agricultural banks and other rural credit institutions such as cooperatives with specific responsibilities to lend to farmers, providing the lending incentive of credit guarantees against loan defaults, and softening terms in cases of agroecological or economic shocks. Finally, attempts were made to make illegal the sort of informal moneylending practices identified by Crow in Bangladesh.

But farmers' need for finance was more complex than could be solved by a cheap loan and benign regulation. The wider financial environment and the delivery systems for credit were also found to be complex. Government intervention in credit, like that in many other markets, was often costly and ineffective. Lending was often politicized and went to the less-poor farmers. Two important structural issues associated with credit arrangements were that lending was often linked to technology that was inappropriate for the poorest, and provision of credit was not accompanied by the provision of other financial services necessary to meet the complex needs of smallholders. Funds lent to farmers were often diverted to other uses, and poor loan-repayment rates contributed to the financial failures and withdrawal of state organizations from credit markets. State failures led to reduced intervention in markets becoming one element of more generalized structural adjustment policies (Poole et al., 2013).

At the same time, international commodity agreements had been established between growers and buyers of commodities like coffee and cocoa which were intended to manage markets and reduce risks incurred throughout the chains through the management of quotas and the buffer stock programmes. But both the International Coffee Agreement and the International Cocoa Agreement of the 1970s failed in their objectives to stabilize prices (Traoré, 2009). As the agreements broke down, the divergence of prices between newly competing countries resulted in smuggling of commodities and in greater market instability. This caused an increase in and the transfer of price and market risks from parastatal commodity boards to the millions of smallholder producers.

Newer financial movements

Informal financial provision through friends, neighbours, family, small retailers and traders, and, increasingly, remittances from migrants, has been ever-present. From the 1980s, other finance movements emerged. Donors and NGOs seeking to fill the finance gap left by the state addressed the needs of the poor through more carefully designed programmes of microcredit provision. Loan performance was monitored, repayments were small and regular, the savings function was linked to lending and terms were enforced. Group lending – or rather lending to individuals organized in groups – proved to be a major innovation, avoiding problems of collateral. Collective responsibility was a means of utilizing peer pressure to effect joint liability and overcome the problems of information asymmetries and the transaction costs of reaching many small borrowers. The Grameen Bank in Bangladesh, which started in 1976, has been the most influential and successful organization in microcredit and, subsequently, the provision of broader and more flexible financial services – microfinance. Latterly, the exploitation of mobile telephony to manage payments between financial institutions and among individuals has become widespread, particularly in Africa.

The microfinance movement has not resolved all the problems of lending and the provision of other financial services to small farmers. Formal systems still suffer from the weaknesses of high cost, low return, limited reach, and lack of knowledge about farmers' requirements. Specialized agricultural banks such as NATSAVE in Zambia continue to operate, more or less successfully. NGOs have found new ways to channel donor finance to the poorest. Other credit institutions such as savings and credit cooperative organizations act as conduits for public finance directed towards the poorest. Thus, there is a wide range of approaches, as summarized in the typology of financial institutions shown in Figure 3.1.

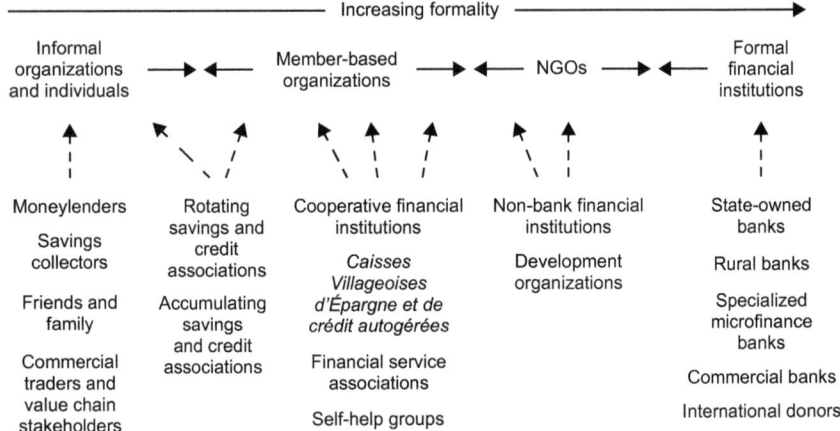

Figure 3.1 Spectrum of financial services providers
Source: adapted from Helms (2006)

Finance and transaction costs

As we noted in Chapter 2, transaction cost economics (TCE) is an approach to the conceptualization and analysis of institutional arrangements – or the diverse forms of contracting governing economic exchange – that evolve (or are chosen) in order to reduce transaction costs. Farmers – and most other business people – do not act to minimize costs but to maximize revenue subject to costs (i.e. profit), including a risk threshold. This risk element is embedded within the calculations and strategies of smallholder farmers and is hugely important in making their production and marketing decisions – and in seeking the finance that is often necessary to take advantage of business opportunities.

Information challenges in financial services provision

Since the financial crash of the mid-2000s, the financial services industry in advanced economies has been criticized for many failures, including self-serving, irresponsible lending and exploitation of borrowers. In developing economies, it is understandable that farmers are averse to borrowing when the risks may lead to severe indebtedness, bonded labour which is akin to slavery, and high rates of suicide. A certain level of risk aversion is inevitable and rational. Costs of poor services are also borne by financial intermediaries and other providers who have lost capital through ill-advised lending and irrecoverable loan portfolios.

It is pertinent to consider the fundamental challenges in delivering financial services which are rooted in common behaviour patterns.

Two specific examples of behavioural problems are rooted in asymmetric information, where one party to a transaction has more information about market characteristics than another. Moral hazard is the first, for which a definition (derived from Investopedia, undated) is: *the risk borne by one party to a transaction that another party to the transaction has not entered into the contract in good faith, and/or has provided misleading information about its assets, liabilities or credit capacity, and therefore may take unusual risks in an attempt to earn a profit.* In a lending context, moral hazard refers to a tendency by borrowers who have secured finance to fail to protect themselves against risk, trusting rather in the services and possible payouts than prudent risk-reducing behaviour. Similar phenomena are commonly observed in an insurance context.

A second, related phenomenon is adverse selection, which occurs where there is a tendency for one party to a transaction, usually the buyer of goods or services, to be among those people who engage in less prudent behaviour. Asymmetric information is implicated again, where *one party usually has a greater awareness and understanding of the goods or services being bought and sold,* and can exploit the other party. For example, those seeking financial services may be those potential borrowers who do not take sensible precautions against risks – or the suppliers of poor-quality goods and services where the quality cannot be easily verified, at least until after the transaction is completed.

Thus, these problems where one party to the transaction has better knowledge about the goods, services and likely reactions of market participants, can be considered information failures, and give rise to behavioural problems, essentially of cheating. Transaction-cost economics focuses on these problems of information, uncertainty, and behavioural patterns in markets.

Principal-agent terminology can also illuminate contractual difficulties. Information and behavioural problems afflict the less well-informed partner, who is often the principal, in seeking reliable buyers and sellers, trustworthy recipients, and sustainable businesses (i.e. agents). Assessing creditworthiness, monitoring financial relationships, ensuring compliance with rules, mediating and sanctioning cases of failure and default are all costly business activities in terms of time, management worry, and actual financial expenditure. Thus, firms incur transaction costs of mitigating the uncertainty created by information and behavioural problems, which eat away at a firm's profitability. The total of normal costs of production or service provision plus transaction costs can be so high, as is often the case when transacting with many small-scale enterprises, that transactions do not happen. This is market failure. Institutional economics, of which transaction-cost economics is a branch, attaches importance to the provision of information and the formulation of rules and norms of behaviour – or institutions – which reduce incentives for behavioural problems and so lower transaction costs.

The informational problems in markets that apply clearly to the provision of financial services have the following implications:

- In a world of imperfect information, and thus increasing uncertainty, transaction costs that individuals and firms incur increase, driving a wedge between total costs and profits.

SMALLHOLDER AGRICULTURE AND MARKET PARTICIPATION

- At some high level of uncertainty, total costs rise to a level too high for transactions to take place and the market fails.
- The provision of information and development of market institutions reduce transaction costs, shifting total costs from *ab* to *cd* (see Figure 3.2), and enable more business to take place.
- Institutions can be strengthened by formulating product standards, specifying clear contractual arrangements, informally by increasing trust through repeated successful dealings, and formally by strengthening judicial procedures.

Figure 3.2 depicts the impact of informational problems on costs and profits. As uncertainty increases, transaction costs rise, and drive a wedge into the margin between the cost of production and the sales price of a good or service. In effect, the total costs of production and transaction costs rise, eating into the real profits of the seller. At some high level of uncertainty, the total costs of producing and transacting business exceed the benefit or profits, and it is no longer worth the seller engaging in business. Institutional mechanisms such as credit guarantees, clear contracts, and sound enforcement practices which provide information or reduce risk can lower transaction costs. Such mechanisms effect a shift in the cost curve from *ab* to *cd* (see Figure 3.2) and make business profitable beyond the earlier point of uncertainty and market failure. In short, reducing uncertainty increases the likelihood of successful buying and selling.

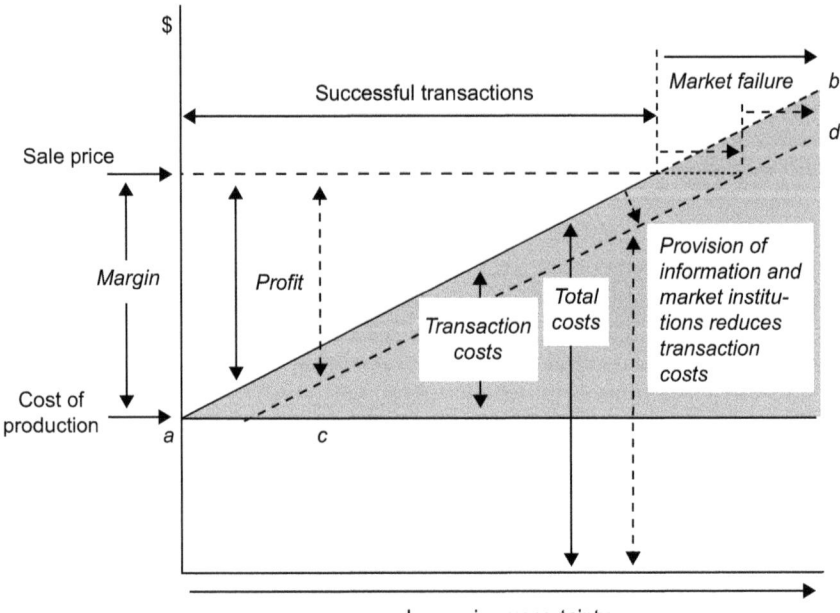

Figure 3.2 Depicting transaction costs
Source: author

As Douglas North said:

> Institutions exist to reduce the uncertainties involved in human interaction ... The costliness of information is the key to the costliness of transacting, which consists of the costs of measuring the valuable attributes of what is being exchanged and the cost of protecting rights and policing and enforcing agreements. These measurement and enforcement costs are the sources of social, political, and economic institutions (North, 1990: 25–7).

Challenges for rural enterprises

Commonly, there is failure in financial markets to deliver services to poor people for good reasons of information, scale economies, and institutional barriers:

- The costs of acquiring information about the creditworthiness of borrowers are high.
- Loans are likely to be small, costly to administer, and relatively unprofitable for the lender.
- Risks of low repayment are high because of the uncertain nature of farming and other rural enterprises.
- Collateral is often absent where enterprises are small and individuals have limited capital and insecure property rights to resources such as land.
- The financial culture and services are rudimentary, and loans are not often linked with savings and risk management.

Like other natural resources-based industries, farming generally is a risky enterprise, and rural households usually maintain a diverse range of livelihood activities as a risk management strategy. Besides crop, livestock, and tree production, rural people engage in off-farm employment and migration, both of which, together with savings, can provide resources for household expenses and capital for farming enterprises. Development of rural enterprises often requires external finance and can be achieved through different means, such as increasing farm productivity by intensification of technology, expansion of scale of production, and diversification of farm activities.

Farming presents particular financing problems. Farming is subject to 'asynchronous' costs and returns: for example seasonality in, and lags between, the need for inputs and the sale of outputs, giving rise to highly significant gaps between incurring costs and receiving returns. For perennial activities such as tree crops and large livestock, this gap is more critical than for annual crops and small livestock, which turn over within a year. Farmers are also subject to individual or idiosyncratic risks and shocks, and covariant risks, which are those shared by a group, such as drought and flood. Farming households also have both ongoing livelihood expenses such as consumption requirements – few rural households are self-sufficient in food – rents and school fees. Both farming and off-farm enterprises often require lumpy expenses, such as investment in new technology – tools and machinery – and new productive assets such as land, livestock, and trees.

Risk in agriculture and the rural economy

Hardaker et al. (2004) have written extensively about risk in agriculture. Drawing on this work, AAACP (All African Caribbean Pacific Agricultural Commodities Programme) research reports on risk management in Small Island Developing States in the Caribbean and the Pacific (Angelucci and Conforti, 2009), and in Fiji specifically (Veit, 2009), refer to a series of different types of risk that affect farming households. These can be summarized as:

- price risk – a sudden unanticipated change in input and/or output prices;
- production or yield risks – arising from natural hazards such as sudden agro-climatic shocks and disease that affect the quantity and/or quality of output of crops, tree products, livestock;
- asset risks – associated with theft, fire, and other types of loss or damage to equipment, buildings, and other assets;
- institutional risk – resulting from changes in national and international policies that affect imports, exports, and standards;
- market-structure risk – changes in the concentration of market power along a value chain, particularly associated with the strategies of large buyers and sellers;
- financial risk – arising from unexpected changes in the cost of capital, exchange-rate fluctuations, or disruptions in the ability to access credit and/or equity losses;
- human or personal risk – idiosyncratic or individual risk arising from household illness, injury, or death.

It is also timely to consider climate-change effects. While evidence for the attribution of climate change to anthropogenic carbon emissions causing global warming is almost overwhelming, there are still elements of uncertainty (uncertain knowledge about climate change) and risk (uncertain consequences about climate change which affect farmers' decision-making processes, in terms of the scale of loss and the probability of loss occurring). It seems inevitable that farmers will increasingly have to incur extra costs to adapt to, and mitigate the effects of, climate change. The precise locational effects are much less certain and create further informational problems for financial services providers.

The financial challenges arising from rural enterprises such as farming do not only affect individual farmers. When farmers organize collectively, for example to form cooperatives, then the same challenges are repeated but are magnified. These are not only a matter of scale, arising from the sum of many individuals' risks, but are also due to an increase in the scope of risks from the more complex form of business organization that a cooperative represents. The multiple functions that were mentioned in Chapter 1 give rise to risks associated with input supplies, accumulation of farmers' products of crops and livestock, plus, often, processing, manufacturing, and sale of transformed products. To this can be added the provision of technical services as well as financial services to members, and finally the need for more sophisticated forms of business governance – often participatory – and executive management.

Financial innovations for farming
Supply of financial services

Despite – or perhaps because of – these challenges, innovations in financial services and in forms of delivery are taking place, with different sources of capital becoming available:

Leveraging equity. Pretes' (2002) work is not so new, but he publicized the Village Enterprise Fund, a US NGO in East Africa, using equity-based microfinance to support small business start-ups and facilitate the growth stage of the business. He showed how start-up grants and equity finance are useful and appropriate additions to the more common loan-based approaches.

Using remittances. According to Sutherland (2013), recent data show that remittances from migrants' earnings constitute a major and irreplaceable flow of resources approaching $500 bn per annum, with around 250 million migrants financially supporting 1 billion people in countries of origin.

Finance and risk management. Angelucci and Conforti (2010) show how finance is more than credit: financial management also involves identification and management of risk. They showed how the provision of necessary insurance can be hindered by a lack of credit and underinvestment in productivity improvements. Also, market context matters: for Small Island Developing States (SIDS) lack of demand due to small farming populations constrains the development of appropriate insurance products – work to which we will return shortly.

Value chain links. Financing through supply chain contracts and the growing focus on value chain management has opened new possibilities. As illustrated by Miller and Jones (2010), value chain financing can be focused on chain activities and linkages and complement conventional lending by being embedded within other services, reflecting stakeholder participation in terms of shared risks and returns. Since 2008 the United Nations World Food Programme (WFP) Purchase for Progress (P4P) initiative has aimed to provide a guaranteed market to smallholder farmers and at the same time to link farmers with commercial organizations that can provide the necessary supports and services, including information technologies, to develop viable agribusinesses (World Food Programme, 2016). Forward delivery contracts from WFP reduce the risks to farmers who might otherwise avoid engagement with markets. Governments often participate by creating a conducive business environment. WFP build and coordinate strategic partnerships with all value chain partners, including firms and other organizations such as AGRA, Bayer, GrowAfrica, the International Finance Corporation, Rabobank, Syngenta, WFP, and Yara International.

Private-sector contractual finance. Experience in Nigeria shows that the model of linking value chain functions and partners can be formalized. McMahon and Harrison (2014) explain how a commercial bank has engaged in a smallholder financing scheme in Nigeria. The bank provides the agricultural system with access to the finance and value chain relationships that are needed for increased productivity. The key multi-actor partnerships are value chain linkages with produce buyers, input providers, and smallholder cooperatives: the bank provides credit to farmers through the cooperatives of which they are members; and an input supplier provides training and production services including land preparation and high-yield seeds, technical advice, and harvesting services. An essential link is that a milling firm provides a guaranteed market outlet. The opportunity for the bank is to develop its own customer base among the millions of rural people. The cooperative is a fundamental link: the risk of farmers side-selling produce to other traders is minimized by linking services to farmer members within the approved cooperative organization. Finally, compulsory crop insurance covers the cooperatives, members, and hence the commercial partners, from the risk of natural disasters.

Social capital development. Sidibé et al. (2014) showed how finance in Mali has a wider social impact: the harvesting, processing, and marketing of the West African shea industry is primarily in the hands of women, often individually but also through collective organizations. An increase in working capital enabled a cooperative to upgrade quality-control processes and expand sourcing from members and also, increasingly, from non-members and traders. At the same time, the business relationships between the cooperative and local suppliers, and with financial services providers, were strengthened and formalized, showing how access to external finance led to changes in the cooperative's institutional structure and in arrangements with both internal and external stakeholders. Similar results were found in the shea industry in neighbouring Ghana, where value chain organization enabled new formal contractual arrangements, shared investments, and quality improvements (Kent et al., 2014).

Loan guarantees. In their review of public–private sector partnerships in Africa, Poulton (2009) and Poulton and Macartney (2012) explored how different forms of delivery have potential to alleviate binding constraints to investment in agriculture by the private sector. Credit or loan guarantees – which constitute insurance for financial services providers – are one such mechanism that is becoming more widespread. A loan guarantee fund allows a financial institution such as a small credit union to lend to riskier clients on the basis that the risk of default is borne by a larger financial provider, say an investment bank or public-sector authority, which assumes the debt obligation in case of default. The latter allocates a sum of money to the scheme and lends to interested small-scale finance institutions who may have better information about borrowers but still cannot absorb the risk of default.

Linked services. taking a livelihood-capital perspective on finance, Donovan and Poole (2014) argued that livelihood assets show complementarities and trade-offs, and illustrate how financial capital underpins most or all of the other forms of asset of coffee-growing smallholders in Nicaragua:

> Financial capital is more than income or credit arrangements. Working financial capital underpins investment in other livelihood assets, particularly natural and physical, such as fertilizer (for maintaining natural capital) and agricultural equipment and roofing (for physical capital). It is also an important entitlement mechanism to meet general household expenses and other human capital-building pathways such as educational expenses for children. Thus financial capital has two important characteristics: it is a means to an end rather than an end in itself; and it is fungible: actually it is a means to various ends. But while the provision of credit is of primary importance, it is not a panacea (Donovan and Poole, 2014: 12).

Other examples of public–private partnerships. The UK Enterprise Finance Guarantee (EFG) is a loan guarantee scheme to facilitate lending to viable businesses that have been turned down for a normal commercial loan because of a lack of security or proven track record. The delivery of EFG, including all lending decisions, is fully delegated to the lender, while the government provides a guarantee to the lender (see British Business Bank, undated). Under the Bank of England's scheme, Funding for Lending, banks and building societies may borrow from the Bank of England at cheaper than market rates for up to four years (UK Government, 2015). By lowering interest rates, financial institutions increase access to credit and thus lending to businesses.

Novel sources of finance. Finally, new sources of finance are opening up: for example, grants, loans, and equity from non-traditional sources. While these may include remittances, they also include funding from foundations, philanthropic capital, crowdfunding, impact investment, and 'patient' capital.

Credit – and other services – to farmers

Miller and Jones (2010) have explained their understanding of what is meant by a value chain approach to financing agricultural enterprises. They make an important qualification about the potential – the limited potential – of better financing to farmers. Finance alone is not enough, as we have seen with collective organizations. Complementary or supporting services are necessary to leverage the impact of efficient financing:

> Even though finance is often a necessary requirement in successful value chains, finance alone is generally not sufficient. The business development

services associated with value chains or market development may be more important to success than the financial inputs. Being aware of the gaps and opportunities in a value chain, and promoting partnerships and ways to address hurdles that go beyond the capacity of the financial institution to resolve can improve the results of the value chain partners and those who finance them (Miller and Jones, 2010: 155).

New financing initiatives are needed, and new opportunities are continuing to open up. In AAACP reports on the Zambian cassava sector, Poole identified the diverse challenges (Poole, 2010; Poole et al., 2010):

> Appropriate lending mechanisms to large private sector firms and to smaller-scale processors are a challenge when lenders consider the enterprise to be high risk and low potential reward. Innovative systems of financing need to be employed to channel development funds to lending organizations through competitive tendering. Firms and organizations within the sector can engage in competitive tendering for grants and loans for enterprise development maybe in partnership with supply chain stakeholders ... New funding mechanisms are also contingent on two other elements: adoption by producer organizations of a business structure that exploits the potential of new-generation cooperative organization; and innovative means of leveraging private-sector investment into collective (probably community-based) organizations. Group lending offers particularly good prospects for generating rural enterprises. Such an approach is a means of capitalizing forms of collective enterprise for rural processing based on rural organizations which are most likely to be community-based, or founded around some other collective entity or ideal like local faith organizations. The development of farmer organizations will continue to depend on external players for investment, equity, management and technological inputs. What is necessary is a realistic timeframe. Achieving sustainability is a very long term process: if 'economic sustainability', or organizational maturity means 'independence of outside agencies', then considering the common trajectory of farmer collectives, such initiatives may take years or decades to reach maturity ... (Poole et al., 2010: 22–3).

The need for new forms of financial delivery and the lack of interest from the private sector so far suggests that the conditions of market failure are present to justify carefully designed intervention and financial innovation. There is scope for further research in delivery mechanisms for this level of microfinance: new knowledge and evidence is needed to design appropriate financing mechanism, particularly for delivery of small-scale funds to grassroots organizations: micro-funding may be up to $10000 for infrastructure for an individual processing plant. Private-sector business-service firms (such as accountants) can be invited by national banks and international financial organizations to design and implement models of competitive tendering and challenge fund approaches for micro-enterprise development. Similarly, private investors can be invited to participate in micro-equity funds willing to invest in such enterprises ... (Poole, 2010: 27).

References

Angelucci, F. and Conforti, P. (2009). *Risk Assessment and Finance in the Fruit and Vegetable Value Chain. Evidence from Small Island Developing States in the Caribbean and the Pacific.* EU-AAACP Paper Series. No. 6. Rome, Food and Agriculture Organization of the United Nations. Retrieved 28 March 2017, from http://www.fao.org/fileadmin/templates/est/AAACP/pacific/FAO_AAACP_Paper_Series_No_6_1_.pdf.

Angelucci, F. and Conforti, P. (2010). Risk management and finance along value chains of Small Island Developing States: evidence from the Caribbean and the Pacific. *Food Policy* **35**(6): 565–575 <https://doi.org/10.1016/j.foodpol.2010.07.001>.

British Business Bank (undated). Understanding the Enterprise Finance Guarantee. Retrieved 9 May 2017 from http://british-business-bank.co.uk/ourpartners/supporting-business-loans-enterprise-finance-guarantee/understanding-enterprise-finance-guarantee/.

Crow, B. (1989). Plain tales from the rice trade: indications of vertical integration in foodgrain markets in Bangladesh. *Journal of Peasant Studies* **16**(2): 198–229 <http://dx.doi.org/10.1080/03066158908438390>.

Donovan, J. and Poole, N.D. (2014). Changing asset endowments and smallholder participation in higher value markets: evidence from certified coffee producers in Nicaragua. *Food Policy* **44**: 1–13 <http://dx.doi.org/10.1016/j.foodpol.2013.09.010>.

Hardaker, J.B., Huirne, R.B.M., Anderson, J.R. and Lien, G. (2004). *Coping with Risk in Agriculture.* Wallingford, UK, CAB International.

Helms, B. (2006). *Access For All: Building inclusive financial systems.* Washington DC, Consultative Group to Assist the Poor, World Bank. Retrieved 09 July 2014, from http://www.cgap.org/sites/default/files/CGAP-Access-for-All-Jan-2006.pdf.

Holtzman, J.S. (1989). Maddening myths of agricultural marketing in developing countries. *Journal of International Food and Agribusiness Marketing* **1**(2): 55–62 <http://dx.doi.org/10.1300/J047v01n02_05>.

Investopedia (undated). Moral hazard. Retrieved 9 May 2017, from http://www.investopedia.com/terms/m/moralhazard.asp.

Kent, R., Bakaweri, C. and Poole, N.D. (2014). Facilitating entry into shea processing: a study of two interventions in northern Ghana. *Food Chain* **4**(3): 209–224 <http://dx.doi.org/10.3362/2046-1887.2014.022>.

McMahon, J. and Harrison, T. (2014). *Collaborating for Smallholder Finance: How is Stanbic closing the loop?* Inclusive Business in Practice – Case studies from the Business Innovation Facility portfolio, Inclusive Business Hub Team. Retrieved 28 March 2017, from http://api.ning.com/files/poTC6m2b82oCVTyU-6XyzAWFttZzcEg98SWPbveUwwcfxHM-Z4PVVjZNXnvHiyl3V4OfC4BYYLNMy80mcNyB7dEGItSoC73O/Deepdive_Stanbic_HUB.pdf.

Miller, C. and Jones, L. (2010). *Agricultural Value Chain Finance: Tools and methods*. Rome and Rugby, UK, FAO and Practical Action Publishing.

North, D.C. (1990). *Institutions, Institutional Change and Economic Performance*. Cambridge, Cambridge University Press.

Poole, N.D. (2010). *Zambia Cassava Sector Policy – Recommendations in Support of Strategy Implementation*. EU-AAACP Paper Series. No. 16. Rome, Food and Agriculture Organization of the United Nations. Retrieved 28 March 2017, from http://www.fao.org/fileadmin/templates/est/AAACP/pacific/07_FAO_AAACP_Paper_Series16_Recommendations_Zambia_Cassava_Strat.pdf.

Poole, N.D., Chitundu, M., Msoni, R. and Tembo, I. (2010). *Constraints to Smallholder Participation in Cassava Value Chain Development in Zambia*. EU-AAACP Paper Series. No. 15. Rome, Food and Agriculture Organization of the United Nations. Retrieved 28 March 2017, from http://www.fao.org/fileadmin/templates/est/AAACP/eastafrica/FAO_AAACP_Paper_Series_No_15_Constraints_to_smallholder_participaÃ__1_.pdf.

Poole, N.D., Chitundu, M. and Msoni, R. (2013). Commercialisation: a meta-approach for agricultural development among smallholder farmers in Africa? *Food Policy* **41**(August): 155–165 <https://doi.org/10.1016/j.foodpol.2013.05.010>.

Poulton, C. (2009). *An Assessment of Alternative Mechanisms for Leveraging Private Sector Involvement in Poorly Functioning Value Chains*. EU-AAACP Paper Series. No. 8. Rome, Food and Agriculture Organization of the United Nations. Retrieved 28 March 2017, from http://www.fao.org/fileadmin/templates/est/AAACP/eastafrica/FAO_AAACP_Paper_Series_No_8_1_.pdf.

Poulton, C. and Macartney, J. (2012). Can public–private partnerships leverage private investment in agricultural value chains in Africa? A preliminary review. *World Development* **40**(1): 96–109 <https://doi.org/10.1016/j.worlddev.2011.05.017>.

Pretes, M. (2002). Microequity and microfinance. *World Development* **30**(8): 1341–1353 <https://doi.org/10.1016/S0305-750X(02)00044-X>.

Sidibé, A., Vellema, S., Dembélé, F., Témé, B., Yossi, H., Traoré, M. and Kuyper, T.W. (2014). Women, shea, and finance: how institutional practices in a Malian cooperative create development impact. *International Journal of Agricultural Sustainability* **12**(3): 263–275 <http://dx.doi.org/10.1080/14735903.2014.909640>.

Southworth, V.R., Jones, W.O. and Pearson, S.R. (1979). Food crop marketing in Atebubu District, Ghana. *Food Research Institute Studies* **17**(2): 157–195.

Sutherland, P.D. (2013). Migration is development: how migration matters to the post-2015 debate. *Migration and Development* **2**(2): 151–156 <http://dx.doi.org/10.1080/21632324.2013.817763>.

Traoré, D. (2009). *Cocoa and Coffee Value Chains in West and Central Africa: Constraints and options for revenue-raising diversification*. EU-AAACP Paper Series. No. 3. Rome, Food and Agriculture Organization of the United Nations. Retrieved 28 March 2017, from http://www.fao.org/fileadmin/templates/est/AAACP/westafrica/FAO_AAACP_Paper_Series_No_3_1_.pdf.

UK Government (2015). 2010 to 2015 Government Policy: Business Enterprise. Retrieved 9 May 2017 from https://www.gov.uk/government/policies/making-it-easier-to-set-up-and-grow-a-business--6/supporting-pages/getting-banks-lending.

Veit, R. (2009). *Assessing the Viability of Collection Centres for Fruit and Vegetables in Fiji: A value chain approach*. EU-AAACP Paper Series. No. 7. Rome, Food and Agriculture Organization of the United Nations. Retrieved 28 March 2017, from http://www.fao.org/fileadmin/templates/est/AAACP/pacific/FAO_AAACP_Paper_Series_No_7_1_.pdf.

World Food Programme (2016). P4P Overview: Connecting Farmers to Markets. Retrieved 28 March 2017, from https://www.wfp.org/purchase-progress/overview.

CHAPTER 4
Risk management for agricultural smallholders

Agribusiness involves risk. This chapter explores the concept of risk and considers how agricultural smallholders manage in practice the different kinds of risks in production and marketing, and how they cope with price variation. Contracts are one of the most important mechanisms for reducing risk, but the terms of contracts will often depend on the negotiating power of the stronger contracting party. We make the case for governments to be involved in risk management in order to enhance the functioning of markets. The provision of insurance is a common way of supporting businesses, and is increasingly feasible in small-scale agricultural enterprise and marketing. Finally, we comment on risk management challenges in different agricultural sectors.

Keywords: risk, coping strategies, transaction costs, cheating, insurance, contracts, coordination, government

Introduction

The evolution of the food chain from a competitive industry characterised by many participants at all levels to an increasingly integrated system provides a unique risk management opportunity to those who have market power. In the absence of effective intervention by public institutions, highly integrated firms are able to transfer the majority of unacceptable risk to the ends of the chain; in particular, to farmers, ranchers and retail consumers ... While public policy intervention may partially mitigate risk through a variety of programmes and regulations designed to address risk symptoms, past and current policies have generally failed to address a primary cause of the inequity in risk transfer by not ensuring an adequate level of competition throughout the food chain (Swenson, 2000: 65–7).

To begin this chapter with a quotation that criticizes the unfair balance of risk in food chains makes a strong statement about where responsibilities lie for the management of smallholder farmers' problems. The discussion of agricultural value chains has often touched questions related to the risks involved in the transition to a new organization of the agricultural production and distribution sectors. But with the notable exception of Jaffee et al. (2008), few contributions to the literature have been explicitly devoted to the issue of risk management within developing countries' rural agricultural and food chains. Given the adoption of closer forms of value chain integration, simplistic settings that consider natural environmental factors such as weather, pests, and natural disasters as the major source of risk are no longer adequate to represent the economic conditions of farming families in developing countries. More market engagement means that other events are becoming more relevant in determining the overall levels and types of risk to which developing countries' agricultural producers are exposed. These are problems such as contract breaches by a contractual counterpart, disruptions in transportation logistics, unpredictable changes in final demand due to safety concerns of consumers, and the introduction of new standards required to access rich consumer markets.

Some of the initiatives described at the end of the last chapter illustrate how close contractual arrangements can facilitate the entry of farmers into markets. How these arrangements work depends on how they are designed in relation to sharing and reducing the risks that farmers are assume when they engage in markets.

Siegel and Jaffee (2008) have made a valuable contribution to the analysis of the problems of risk assessment and risk management in agri-food chains, with the aim of developing an operational framework to analyse and manage supply-chain risk. They borrowed extensively from the existing literature on supply-chain risk management (SCRM) and on the strand of development literature that has been concerned with the concept of vulnerability.

SCRM sees the value chain as an economic framework or entity within which firms and individuals interact in order to extract a premium over production costs from the sale of a given product. The objective of SCRM analysis is to identify and mitigate the major events that hinder the objective of bringing the 'right products (quantity and quality), in the right amounts, to the right place, at the right time, and at a competitive cost' (Jaffee et al., 2008: 5). The concept of vulnerability underlies risk. Vulnerability is defined as the combined result of the probability of an event occurring and of the economic damage that may derive from such an event. In SCRM, vulnerability is not an individual or household characteristic but is taken to be the vulnerability of the chain itself; that is, the possibility that the chain might *fail to bring the right product in the right quantities to the right place at the right time and at a competitive cost*. The management of risk in the chain is seen from the perspective of a fully vertically integrated enterprise, concerned with the risk of the disruption of the chain, with limited or no attention to the distribution of risk among chain participants.

The issues discussed in this chapter are the same ones that inform SCRM, but less at the level of the chain as a whole than at the level of chain participants. The questions are: how to define and identify the risks that are relevant to small-scale farmers and entrepreneurs, how to analyse the available or potential risk management tools, and how to suggest possible policy responses. First, the focus here will be on 'price' or 'market' risk for agricultural producers, making only passing reference, when deemed necessary, to the problem of managing 'production' risk, which has been more extensively covered in the literature.

Second, the problem of risk exposure and risk management will be considered mainly from a 'micro' perspective, that of the welfare of poor rural households.

Here we will consider how risks can be identified and managed in rural value chains, and who is affected in chains such as those for farm products. The chapter draws substantially on the AAACP papers by Cafiero (2008) and Conforti (2009).

Assessing risk in rural value chains

There are three conceptually distinct processes in the analysis of the welfare consequences of residual risk exposure and of the potential welfare-improving character of new public policies. If we take as given the organizational structure of the value chain, the steps are: (a) identification of the subject exposed to risk; (b) definition of the *ex ante* condition of risk exposure; and (c) assessment of the potential welfare consequences of residual risk exposure, once the set of possible private risk management actions by the chain participant have been duly considered. Each of these analytic processes presents some difficulties, which will be briefly described below.

The fundamentals: identifying and mapping risks in the chain

Siegel and Jaffee (2008) list all possible sources of risk potentially affecting agricultural value chains. They identify seven categories: weather-related risks, natural disasters (including extreme weather events), biology and environmental risks, market-related risks, policy and institutional risks, logistics and infrastructural risks, and management and operational risks. They then discuss each category in some detail, from the perspective of the whole value chain.

This approach provides interesting information. However, listing the potential sources of risk does not provide practical suggestions on which kinds of policies might be more suitable to the objective of protecting the interests of one or some of the various value chain participants. This is because not all of the listed sources of risk will have the same relevance or the same kind of implications for all participants.

Any shock that causes risk may 'hit' one or more of the relevant links that form the value chain. The consequences of the shock may then be partly or entirely transferred to other agents within the supply chain. This depends on the kinds of contractual and informal links existing between the different agents. For an informed analysis of the consequences of risk exposure for the various chain participants, each possible source of risk first must thus be traced back to the locus where it 'hits' the various chain functions involved. The next step is to assess the extent of risk transmission occurring at each of the links. Finally comes the question of evaluating how the risk affects the welfare of the different chain participants.

Mapping the particular value chain helps to identify the relationships between the persons or actors in the chain and the functions for which they are responsible. Mapping also highlights the chain mechanisms that link the various functions, and associates the chain functions to the links where the risk is first likely to have an impact.

In mapping, the chain actors are classified in groups, such as households that are poor, middle income, and rich. The chain functions are the traditional ones identified in many chain analyses, namely: input provision, production, trade, processing and manufacturing, retail, and consumption. Not all of these functions will be relevant in all chains. Possible sources of risk are those already identified, such as: weather, pests, and other natural phenomena; idiosyncratic accidents involving injuries; illnesses; and domestic policy, international policy, and generic micro- and macroeconomic shocks.

Before moving to a discussion of risk in actual supply chains, it is useful to understand the process of risk management and the elements of a risk management strategy that may be used to limit the negative consequences of risky events on vulnerable people. Usually, a risk management strategy is the combination of different actions which include a preliminary risk and vulnerability assessment, and subsequent risk management choice, possibly followed by monitoring and re-evaluation of the actions taken.

Risk assessment

Risk assessment implies identification of the possible risk-generating events, quantification of the impact that each will have in terms of financial losses, and association of a probability value to each of them. Even though precise quantification of the extent of losses and the probability of the risk occurring is virtually impossible, to have even a rough idea of their range can be extremely important in defining the optimal risk management strategy.

One informative practice, common in financial investment and risk management, is the process of risk layering, in which a probability distribution of potential losses is formed. When the only dimension being considered is the entity of the financial loss, irrespective of the event that might cause it, the distribution will have the shape depicted in Figure 4.1. It can be seen that a proportion of the risks involves smaller financial losses with a high probability of occurring; usually these losses are absorbed by the person or firm affected. This is the layer of 'retained' risk, or the 'retention' layer. Other layers of risk are identified by choosing the levels of financial losses that separate the risk-retention layer from the risk-transfer layer and this latter from the so-called tail risk. This practice is useful in informing the process of risk management through insurance, in that it allows for the identification and handling of different types of risk:

- the risk-retention layer corresponding to the small losses that the individual or enterprise is willing to retain and cope with by bearing the cost through managing their own resources and strategies;
- the part of the intermediate financial losses that the enterprise might want to insure against (the insurance layer); and

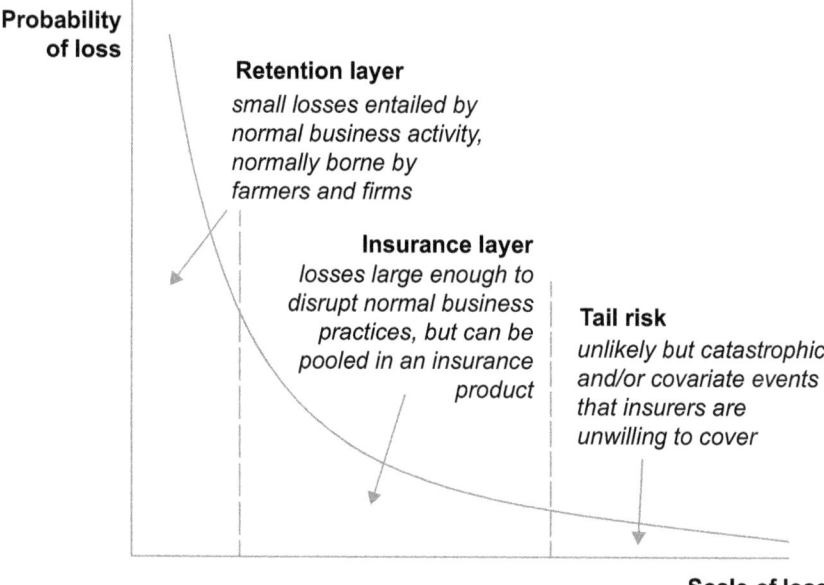

Figure 4.1 Risk layering
Source: author

- the tail risk of unlikely but catastrophic losses which is the part that will not be covered through formal insurance.

Despite its usefulness, the process of risk layering is rarely emphasized in the agricultural economics literature, probably because of the difficulties posed by the process of quantifying the extent of losses and the associated probabilities, and also because of the limited diffusion of commercial insurance in agriculture beyond advanced-economy countries. What makes the risk layering process useful, however, is that it forces reflection on the cost/benefit ratio associated with transferring the risk. In principle, any risk, no matter how frequent, could be transferred but that would imply an unacceptably high insurance cost. In addition, the welfare effect of highly frequent, small losses may be limited and lower than the cost of transferring the risk to others, which would imply excessive transaction costs for monitoring and assessing the losses. Usually, self-insurance (a form of risk management) can take care of such events efficiently, and victims of small risks can adapt their livelihood strategies accordingly. However, self-insurance clearly has limitations and more formal provision of insurance is increasingly considered to be valuable. Thus, formal insurance is becoming an important factor in rural development as a way of reducing poor people's vulnerability to external risks and enabling them to escape poverty traps.

Risk layering is a useful exercise, but quantifying losses and probabilities associated with risky events is not easy in developing-country agriculture. A number of interesting insights, however, arise from considering the elements which can affect the size of each layer. The distribution of the layers is primarily determined by the specific risks of each environment; however, the market for services used by farmers and government policies also contribute.

The retention layer. As retention-layer losses are frequent and of limited size, in most rural contexts they are normally addressed through household income diversification and consumption smoothing. Production variability tends to offer a natural hedge to farmers, given the inverse relation observed between prices and yields. Other key elements shaping the ability of farmers to retain losses individually are individuals' access to the agricultural services market, such as credit and finance, storage and transport facilities, and other services, such as extension and technical assistance.

Through unintended consequences, policies also play a major part in shaping the ability of farmers to retain risks. Price controls reduce the automatic hedging offered by the inverse relation with quantities, and redistribute risk rather than reduce volatility. However, where agriculture is widely supported by public resources, as is the case in OECD countries, the risk-retention layer will likely become large, as subsidies and protection will increase farmers' ability and willingness to retain individual risks directly. In developing countries, public support to agriculture is usually more limited, and this tends to reduce the size of the retention layer, and to increase the need to resort to informal mitigation and smoothing mechanisms.

The insurance layer. The insurance-market layer can be small in developing countries: markets tend to be highly incomplete, just as those for other agricultural services are. Agricultural support programmes can crowd out private provision of insurance: subsidies increase the ability to retain risks individually, and this undermines private incentives to purchase insurance. Premiums may be too expensive for poor farmers, greater than the worst possible insurable outcome, and therefore not a prudent strategy. Poverty may also imply a high rate of discount on the future, reducing the willingness to purchase insurances; and lack of understanding about risky events may prevent farmers from making reasoned decisions. On the supply side, high transaction and delivery costs in remote rural areas may result in high premiums, and undermine incentives for insurance firms to operate. Also, controlling contractual conditions may be difficult because of the asymmetry of information: insurance firms may frequently lack the information on the degree of exposure of farmers to different sources of risks, as data are seldom available for remote communities. Actuarial calculations thus become difficult. For this reason, indexed insurances (see below) are seen as a promising alternative: they can overcome the major constraints such as transaction costs, the need for damage assessments, and the associated asymmetric information and moral-hazard problems.

Tail risk. Tail risk is the 'market failure' layer. These risks are highly infrequent and catastrophic events, that are usually not insured by private companies because of their covariate nature and the magnitude of the associated losses. Lack of information reduces the willingness of farmers to insure against unlikely catastrophic events. Hence tail risks usually call for the establishment of public–private partnerships, allowing transfer of risks to a higher level, for example through reinsurance markets. Examples of partnerships are the Turkish Catastrophe Insurance Pool; the Andhra Pradesh microinsurance programme; index-based weather derivative for farmers facing drought in Malawi; and the Caribbean Catastrophic Risk Insurance Facility (CCRIF), co-financed by the World Bank. Farmers often develop expectations of receiving support from governments in case of extreme events such as droughts, floods, or earthquakes. This may dissuade farmers from learning more formally how to manage risk, increase risks of moral hazard and adverse selection, and depress demand for insurance, hence further reducing the size of the insurance market layer.

In practice, intervention in insurance markets takes different forms. Broadly speaking, there are three approaches taken in public-sector intervention in the agricultural – as well as non-agricultural – insurance markets. Firstly, there is a minimal regulatory role which is necessary for enforcing contracts. Secondly, public resources can be deployed to ease the functioning of private insurance markets by contributing reinsurance for covariate risks and coverage for extreme events. Thirdly, public resources can be employed to directly subsidize premiums, either by organizing supplies through state enterprises, or through private insurance companies. These three approaches are not mutually exclusive. Moreover, the distinction between them is often blurred: premium subsidization and financial support to private insurers is often provided in relation to risks which are assumed to be highly covariate and catastrophic, or where insurance knowledge is very limited.

Risk management

Once the extent of risk exposure has been assessed, the next step is to consider what actions can be taken to reduce the losses associated with risk exposure. Discussions start with the special nature of agricultural production, with its strong dependence on biological processes and a high vulnerability to natural phenomena such as weather and pests, as we have noted. The focus is usually on yield or production risk. This approach has several shortcomings, such as neglect of the possibility of a natural 'hedge' or compensation, such as when prices and production are negatively correlated: a low level of agricultural production often signifies high market prices and vice versa. Therefore, through increases in prices, producers might in part be able to compensate for a reduction in production. In such circumstances, producers' total incomes are somewhat equilibrated and welfare losses are limited, given that what ultimately matters is incomes, not simply the level of production.

Coping strategies

If the physical characteristics of production and the institutional context allow farmers to access storage and hedging, even unexpected price variability can be dealt with by farmers directly – and effectively, compared with public attempts to stabilize prices and incomes. The availability of updated and location-specific technical information is another element that helps farmers reduce their risk exposure. For instance, the application of agronomic techniques can reduce the impact of pest attacks, drought, or other sources of yield variability. Therefore, the extent to which agricultural service markets – from credit to transport, storage to extension services – work effectively and meet farmers' demands determines the size of the retention layer. Where service markets are incomplete, or of limited accessibility, farmers are forced to rely on coping strategies and informal methods to smooth their consumption levels.

Examples are crop and income diversification, or informal networks of relatives and friends, or social safety nets, as well as contracts, such as share tenancy or credit contracts, or business contracts along a value chain in the more favourable cases. While some of these strategies are common in all agricultural environments – for instance crop and income diversification – there are risk-avoidance contexts in which they result in the perpetuation of subsistence conditions. Farmers may easily reject specialization and productivity improvements by investment in improved technology such as new varieties of seeds or improved cultivation techniques. Instead they may diversify, or reduce the exposure generated by costly direct or complementary inputs. Hence risk may prevent increases in agricultural productivity, leaving potentially usable resources idle. Such risk can be the origin of poverty traps, and constitute a key constraint to agricultural development and the improvement of living standards of poor households. Essentially the risk is not 'avoided' but becomes another type of risk, but maybe one which the farmer can more easily recognize, if not actually manage.

In principle, three strategies allow management of price risk. These are: (a) directly affecting the distribution of prices, so as to make negative price swings less likely; (b) hedging against predicted negative price fluctuations; and (c) coping with the consequences of 'bad' or low prices. And these actions can be taken either directly by the chain participants, or form part of a government programme intended to provide income protection.

The ability to assess the potential effectiveness of each risk management strategy depends on the causes of price variability in the first place, and the mechanisms through which these variations are transmitted. From the perspective of producers within complex value chains, one crucial element that needs to be considered is the actions that other participants in the chain might engage in. These actions will eventually condition the variability of prices faced by the producer and the producers' ability to mitigate negative fluctuations.

Effects of production, storage, and marketing decisions

The sources of variation that are relevant for agricultural producers depend on whether they make their impact within the 'farm gate' or beyond it. This corresponds to the traditional distinction made between production (or yield) risk and market (or price) risk. Through decisions on the amount being sold or on the timing of sales, the producer can directly affect the price (s)he receives, and therefore production and marketing decisions themselves can be considered mechanisms for risk management.

The timing of marketing decisions is constrained by the need for income for household costs and for working finance. Farmers require credit – or savings – even to bridge the time between sowing and harvests; hence purchasing inputs, even in the absence of unexpected market or environmental swings, requires finance. Crop and plot diversification, savings, storage, and purchase of financial assets, where they are accessible, are common practices that serve the purpose of managing the expected price variability. In the absence of any other source of credit, produce must be sold – so-called distress sales – to meet immediate financial needs. And such sales often occur at a time when market prices are low, because the timing coincides with harvest, and/or because other people are in the same situation, needing to make sales at the same time.

The extent to which the level of production and the price received by the producer are negatively correlated depends on the relative scale of the producer and the size of the market. Usually, for small producers, changes in individual output and product sales are not capable of affecting prices, and in such cases it makes sense to analyse and manage yield risk independently of the price received. The smaller and more local the market, the higher the possibility that the natural hedge can be effective, as quantities and prices are likely to experience greater fluctuations. But even if individual producers have little scope for affecting the price level through management of the quantity marketed, such that production shocks are correlated across producers, individual production and total production will be correlated, and therefore some scope for a natural hedge may exist.

However, local markets may be integrated with larger markets, even international markets, and therefore prices may fluctuate with supply and demand across a wider economic region. In this case, local supply and demand conditions will not change prices in what is, in effect, a much larger market. In short, selling prices will be independent of local supplies, the natural link between low production and high prices is lost, and there will be no natural hedge: farmers may have to sell their limited supplies at prices kept low by the supply and demand conditions in the wider market. Under such circumstances, the production and market risks can both be severe.

Furthermore, in some circumstances, producers as a group might exploit the ability to increase the price they receive by reducing the quantity they market. This is the rationale for the practice of product withdrawal. When demand is inelastic, even a slight decrease in the quantity sold can raise the price and increase overall revenue. For such a strategy to be effective, however, coordination among all producers that supply their product to a given market is crucial. In small markets, again, coordinated market action can be achieved by collective decisions through, for example, cooperatives.

For cartel-like coordinated marketing agreements to be effective at raising prices, there needs to be disciplined producer organization, and for 'cheating' to be detectable and punishable (Figure 4.2 and Box 4.1). Cheating happens because there is a strong incentive for any single producer to sell all of his or her production at higher prices when all others withdraw theirs, stimulating those very same high prices, leading to side-selling. And for a cartel to be able to coordinate producers and marketing behaviour, information about each partner's sales arrangements needs to be available. Often such market discipline is lacking. The international commodity agreements for coffee and cocoa tended to be ineffective for just these reasons.

The type of product has some bearing on the feasibility of coordinating market action. Withdrawal (and destruction) of excess production is an option for perishable products. Easily storable products, such as grains, present another opportunity for producers to alter the distribution of prices through the management of stocks. When technically feasible, the possibility of private storage management is a very powerful means of price-risk management for producers. Thus, unless there are technical conditions that hinder the possibility of storage, the exposure to price risk for producers should always be assessed by taking into account the opportunities for intertemporal arbitrage that can be exploited through storage management. The consequence of this is that, for storable products, pure price risk is rarely

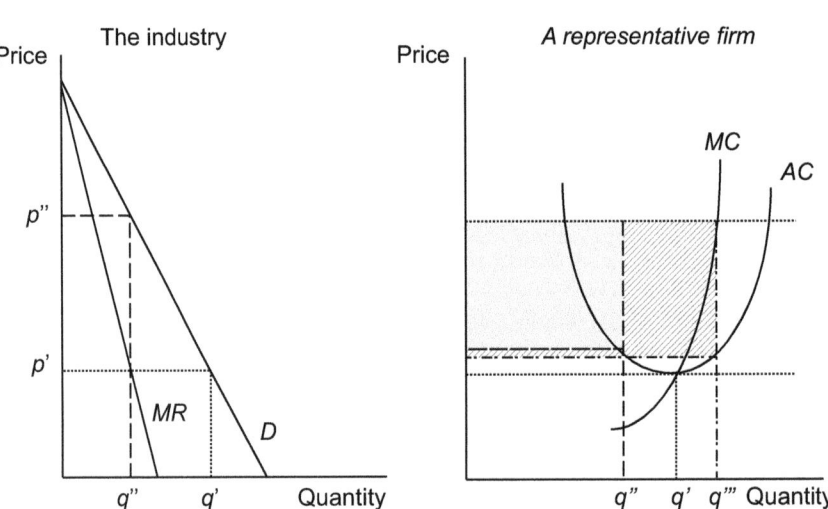

Figure 4.2 The cartel
Source: author

> **Box 4.1 Side-selling: Cheating is the cost of coordinating high prices**
>
> Economic analysis of the cartel problem is a relatively simple way to represent the problem of cheating: sellers try to coordinate output to raise prices, but find that some members of the group choose to increase sales to take advantage of the high prices.
>
> In the left-hand diagram in Figure 4.2 the whole-industry demand curve and marginal revenue curve are D and MR, respectively. The initial market equilibrium occurs at output and price $p'q'$. For the representative firm – or farmer – in the right hand diagram output is set at q', receiving price p', just covering average costs. Now suppose:
>
> - farmers in a cooperative implement quotas which reduce overall industry output to q''
> - the price rises to p'', maximizing joint profits
> - for the typical farm, output quota q'' gives rise to individual profits the pale shaded rectangle in the right-hand diagram at the new price p''
> - but at the new price p'' the individual farm's profit maximizing output is greater, at the level q''' where the marginal cost MC = new price p'' and including the larger diagonally shaded area
> - at this level of output higher than the quota, the farm's profits rise …
> - … but if all firms increase output, industry output increases beyond q'' and prices fall to the original market equilibrium level
> - the profitability of cheating means that it is difficult to sustain high prices – unless other members of the cooperative can detect the source of the increased output and punish the cheat!
>
> *Source*: author

an issue for producers, and any activity directed towards reducing the cost of storage (which comprises both physical and financial components) will always have very strong implications for producers' ability to cope with price risk.

One other aspect that complicates the analysis of price in the presence of storage is active government intervention. Usually, public programmes intended to achieve price stabilization in developing countries are justified as a means to protect consumers. Consumers are numerous and often more politically active, able to bring pressure on governments, in urban areas. Thus, mechanisms such as buffer stocks and export restrictions are put in place for domestically produced foodstuffs with the aim of preventing food prices from rising above a ceiling level. From the producers' perspective, unpredictable changes in policies that directly and indirectly affect commodity markets are perhaps the most relevant source of price risk independent of producers' own market behaviour. In this respect, government lobbying by producer interest groups can be seen as a mechanism for price-risk management. Such advocacy is another potential focus of collective action by producers, and is a form of long-term risk management.

Hedging price variation

When individual producers are confronted with a large market, so that there are no direct ways for them to modify the actual level and distribution of prices, the possibility of hedging against the residual price-risk exposure remains. Hedging is the activity whereby producers enter in a contractual agreement with the objective of countering possible negative variations of the price. To be sure, the most effective way of hedging the price exposure would be that of entering into a contractual agreement directly with the final consumers, who will have the exact opposite stand in terms of price, and therefore would provide the most effective price-risk sharing opportunity. In most cases this is impractical, given the physical and economic separation that exists between producers and consumers. Nevertheless, there are other hedging mechanisms that could be exploited by agricultural producers, which include forward-sale contracts to be stipulated with an intermediary buyer of the product, and price-contingent contracts to be traded on organized exchanges. The possibility of effectively exploiting the hedging potential of formal contracts traded on local or international exchanges depends largely on the institutional settings that allow for their enforcement, and these are limited for developing country producers.

Two aspects of forward contracting deserve to be highlighted. First, in agricultural produce markets, it tends to be that many small producers deal with few economically larger traders, who may exploit local buying power in formulating the terms of the contract: on quantities, quality, delivery and timeliness, and, above all, the price. Second, when a share of local production is sold through forward contracts, the concept of a 'spot' market price loses some of its significance as the link between the wholesale price that forms at the level of final consumption and the farm-gate price is broken. While this might protect producers from the consequences of market events that may occur beyond the farm gate after the contract has been signed, it usually comes at the price of forgoing the benefits of any rises in price.

One other aspect to be considered is that the signing of a forward contract introduces a new type of risk for the producers, namely the possibility that the buyer or client will default on the terms of the contract and fail to take delivery of the product at the specified conditions. It is precisely this risk that increases as smallholder producers enter into new types of contractual arrangement through more sophisticated mechanisms of market coordination. The extent of this client risk depends strongly on the institutions that support exchange, including the prevailing legal system and the ability that producers have to call upon impartial courts to enforce contracts. In short, the level of institutional development influences the transaction costs and the level of risk, as discussed in Chapter 3 and as clearly explained by North, among others (North, 1987, 1990, 1994).

For export commodities, traders of exports originating in developing countries might find opportunities to hedge their selling-price risk exposure through contracts traded on international exchanges. The possibilities for local producers in developing economies depend, once again, on the existence of either formal or informal contractual agreements between producers and traders, on their relative market power, and the complexity of institutional development.

Coping with price variation

The last element of a producer's price-risk management strategy is the possibility of using other mechanisms to cope with the variation of prices, revenues, or returns. Focusing on income risk, the ability to cope depends essentially on the level of reserve assets such as livestock and machinery, or remittances and other forms of reciprocal exchange, or the ability to draw on personal savings. In other words, the cost of any activity intended to cope with monetary income shocks is given by the opportunity cost of personal savings and/or by the effective interest rate for credit.

Problems arise when access to credit is limited or absent. A full understanding of the role of coping strategies in such conditions requires analysis from the overall household perspective. It has long been established that lack of access to credit makes consumption and production decisions interdependent (Singh et al., 1986). Such a relationship may show up in the most pernicious way following income shocks, and the consequences can be dramatic, jeopardizing the future ability of the household to escape poverty, as when coping strategies involve the mobilization of the productive assets mentioned above, or reducing household expenses and investments in human capital (Dercon, 2005).

In this context, the numerous and growing positive experiences of the microfinance initiatives, and the development of analogous microinsurance programmes, are significant developments in financial markets. Such initiatives lessen the link between income shocks and the long-run asset status of a rural household. In other words, even though it might be described as a means to promote investments in productive activities, the value of microcredit is that it allows smallholder farmers to buffer consumption shocks (Johnston and Morduch, 2008).

Risk management and the role of contracts

The emergence of institutionally complex value chains in the markets for agricultural and food products poses a big challenge to defenders of free markets. These arrangements demonstrate the coordination and information problems that prevent the trade of agricultural products from being efficiently regulated by free market forces alone. With reference to the risk implications, unfettered markets become a means through which those who can enjoy greater bargaining power can transfer the majority of unacceptable risk onto others, rather than a mechanism through which the socially optimal distribution of risk is achieved. It turns out that, in business, it is not just prices that matter. Rules and regulations, both formal and informal matter, and interpersonal relationships are included: as Fafchamps and Minten (1999) found in Madagascar, relationships are important because they provide safeguards where business practices are unsophisticated. However, without supporting institutions in the business environment, 'the free market remains nothing but a flea market' (Fafchamps and Minten, 1999: 31).

Quotes like the one reported at the beginning of this chapter might convey the impression that the ability to extract rents associated with market power can also be exploited to transfer more than the 'fair' share of risk. Nevertheless, in discussing risk management from a value chain perspective, it is imperative to analyse the competitive structure at each and every link of the chain and understand how the real conditions of the transaction allow one of the two parties to transfer risk to the other. About 250 years ago, the Scottish economist Adam Smith observed the problem of market power very clearly, and he could equally have been describing farmer cooperatives or traders. Like Fafchamps and Minten, he saw that those without market power were frequently disadvantaged but, unlike them, he envisaged no role for direct government intervention:

> People of the same trade seldom meet together, even for merriment and diversion, but the conversation ends in a conspiracy against the public, or in some contrivance to raise prices. It is impossible indeed to prevent such meetings, by any law which either could be executed, or would be consistent with liberty and justice. But though the law cannot hinder people of the same trade from sometimes assembling together, it ought to do nothing to facilitate such assemblies, much less to render them necessary (Smith, 1776: 232).

Market power usually plays a role in the contract-bargaining process that leads to the definition of the contractual terms. Usually the smaller, weaker party to a contract is forced to accept less advantageous conditions. Yet there may be cases in which the presence of risk offsets the effects of market power, such that it works to the benefit of the competitively weaker party. For example, procurement contracts by the vegetable processing industry, or by the organized retail sector for fresh fruits and vegetables, lead to a sharing of responsibility for some of the production risks in order to avoid disruptions in the continuity of processing and sales.

This rebalancing of power can be seen in the forward sales that are common in the fresh-fruit sector of Mediterranean countries and elsewhere: the buyer (usually a wholesale trader) pre-purchases a crop – often a tree crop such as citrus – before harvest, thereby absorbing part of the yield risk, in order to ensure the procurement of the product (Box 4.2). Commonly, such contracts stipulate the lump sum to be paid to the producer for the entire production of a given field, usually at the end of the winter and right after the pruning of the trees, when only a rough estimate of total production can be made, based on the appearance of the flower buds. In this way, producers transfer the residual yield risk to the traders.

These cases demonstrate that, in specific cases, risk along the value chain is better handled through complex contractual agreements than simply through sales. However, just as smallholder farmers are unlikely to choose contractual arrangements to minimize transaction costs alone, so also to base their contract choice exclusively on risk reduction is probably an unwise strategy.

> **Box 4.2 Formal contracts: Risk sharing between farmers and traders in Spain and Ghana**
>
> Can the introduction of standard seller–buyer contracts facilitate producers' marketing decisions and reduce uncertainty, thereby lowering transaction costs for Spanish mandarin and orange producers? Research showed how producer behaviour was found to vary and how producers could be grouped according to their different production and marketing orientations. Regarding sales decisions, marketing factors in addition to the negotiated price were found to be important determinants of the terms of the transaction. Most of all, uncertainty about payment was a major preoccupation of small-scale producers, more so than prices, and contracts were considered to be one way to mitigate this form of market risk.
>
> Marketing institutions are less developed in sub-Saharan Africa: producers often experience weak bargaining relationships with traders because they do not have access to information on prices, demand conditions, or alternative marketing channels. Most of all, farmers lack the ability to enforce verbally agreed terms of exchange with traders. Farmers may also renege on agreements, to the detriment of small traders. Such contractual inefficiencies reduce the performance of the market system, with multiple consequences: there are unexploited market opportunities, in-field and post-harvest losses, seasonal gluts of produce, poor quality control, inequitable returns to producers, unsatisfied consumer demand, and reduced multiplier effects in the rest of the local economy.
>
> A study of vegetable marketing in Ghana offered ideas for how the development of contractual institutions could enhance market coordination: closer coordination mechanisms to mitigate uncertainty and enhance the building of a clientele, for example through formal contracts, have potential to overcome the pervasive mistrust between farmers and traders and reduce transaction costs.
>
> Written agreements such as standard-form contracts offer three advantages over verbal agreements. First, written specifications establish criteria by which performance can be measured in terms of important variables such as product price; quantity traded; product quality; place of delivery; ownership of packaging materials; and payment date. Changes to these terms will be transparent, and, in the event of unavoidable conflict, resolution mechanisms can be specified.
>
> Second, adoption of written agreements may boost the informal rules of business attitudes and ethics. Moral obligation rather than the force of law may, in time, come to prevail, creating trust between buyers and sellers.
>
> Third, the law is there to help, but expectations of formal legal remedies should not be exaggerated. It may be preferable that unwritten norms and customary laws operate through negotiation rather than the expensive remedial use of contract law through formal legal procedures. Hence the importance of developing moral obligation within the trading culture.
>
> *Source*: summarized from Poole et al. (1998, 2003) and Poole (2000)

Sellers usually make complex calculations which include price, relationship, and risk factors.

The role of governments in risk management

One final issue is the role of government intervention. Of all the reasons why a government might want to intervene in the economy, the provision of some form of insurance is probably the least controversial one. In agriculture, for example, many policies worldwide have been explicitly justified by the desire to reduce risk and stabilize income. The history of stabilization policies in agriculture is long, rich, and diverse, although there still is debate about the overall merit of such policies. With reference to prices, and linked to our discussion of price risk, one accepted conclusion should be that price stabilization per se is not a desirable policy objective. Everything else being equal, an attempt to stabilize prices without an understanding of the fundamental cause for price instability may reduce the natural hedge, resulting in increased risk and a shift of instability from one sector of the economy to another, or from one social group to another, but with no clear net social benefit. Instead, the fundamental source of instability should be addressed.

Nevertheless, direct price stabilization policies and other forms of government intervention in agricultural and food markets are still very common, and they happen to be periodically brought up the policy agenda, following unpredicted swings in the levels of prices. One consideration has to do with the concept of tail risk. Tail risk involves rare but dangerous events, such as natural disasters or extremely adverse economic conditions. In recent years, governments have demonstrated willingness to step in to mitigate financial crises that are a consequence of major tail risks. However, anticipation of this form of intervention will have the consequence that private actors will alter private risk management behaviour and underinvest in private risk-prevention activities.

The second consideration relates to the role that information plays in risk management and in the public-good nature of information. A key step in the design of any risk management strategy is the assessment of the probability of negative events occurring. Information on the frequency with which such events occurred in the past, and on the impact that they had, is crucial to understand what is at stake. Unfortunately, the information may be kept private.

National and international public services play a crucial role in the provision of information. Production risks are increasingly dependent on the impacts of climate change, and the significance of publicly managed early warning systems becomes clear in this context. Projects such as the USAID initiative Famine Early Warning System Network (FEWS NET, undated) provides information on forthcoming likely rainfall and temperature patterns and indicates likely effects on agricultural production in particularly sensitive countries and regions. FEWS NET's estimates of impacts on poor people's livelihoods, and supply and demand impacts on prices and markets and implications for regional and international trade, are the sort of public information that can inform risk management. To collect, validate, certify, and disseminate information that helps in assessing the probability distribution of bad events is perhaps the most important role for public policies in risk management.

With reference to price risk, commodity exchanges typically serve as information brokers for prices of actively traded commodities, and their diffusion in developing countries will certainly help improve the overall market-risk management ability of producers.

In general, the willingness to buy and supply agricultural insurance depends upon farmers' different risk profiles and on the costs at the margin for insurance companies. Insurance is a private service, and excludable: those who pay receive cover; those who don't pay are excluded. As insurance cover is not a public good, the role of the government in the insurance market should be limited to regulation: that is, providing the legal framework for enforcing contracts and ensuring competition. One additional important role for governments, however, is the collection and disclosure of information which can be asymmetric and costly, and is often considered to be a public good. Data on risk exposure and on expected damages are needed on the supply side, in order to perform actuarial calculations about risk, premiums, and likely payouts; and on the demand side, to reduce the lack of specialist knowledge on insurance in general and risk management in particular, and the tendency to underestimate the probability of unlikely events.

Agricultural insurance for agricultural development

Based on a review of experiences with agricultural insurance, this section discusses the extent to which insurance can be employed in developing countries as a market-risk management tool to promote investment and improvements in agricultural productivity. The focus is on the role of public support to small-scale enterprises such as farmers. Successes and failures can be found among experiences entailing different forms and degrees of public intervention, partnership with the private sector, and, to some extent, even within different types of insurance.

Risks generated in agriculture by the interaction with long production cycles imposed by plants and livestock biology are among the justifications which have been given for extensive public support to the sector, along with the importance attached to securing food supplies and access to adequate levels of consumption. In fact, as we have noted, government intervention through domestic supports often has a significant impact on risk. Risk reduction is one of the channels through which subsidies are considered to be 'decoupled' from farmers' production decisions at the margin – for instance those granted under the European Common Agricultural Policy.

Where agriculture is less generously supported than in the OECD, as is the case in most developing countries, farmers deal with risks by resorting to a number of informal and formal strategies and practices allowing them to mitigate risk and prevent excessive income and consumption fluctuations. Despite being effective in many cases, these *ex post* mitigation tools tend to force farmers to adopt a risk-averse attitude, maybe avoiding engagement with markets, hence potentially hindering investment, innovation, and agricultural development, or even perpetuating poverty traps (Dercon, 2005; Barrett, 2008). Enhancing the ability to manage risk can be an important element in the promotion of investment for agricultural development.

Agricultural insurance is not uncommon in advanced economies, and has long been suggested as a risk management and agricultural-development tool in developing countries. Risk-sharing arrangements do not reduce total risk but can reduce the burden on individual farmers in two ways: by transferring the risk to other institutions or organizations who are better able to bear the risk or who are less risk-averse; and by pooling risks across space, crop sectors, and other economic sectors (Hazell et al., 1986: 2). Idiosyncratic and individual risks can be insured against relatively easily. Large-scale phenomena such as drought or floods – the catastrophic tail risks mentioned above – are more difficult to insure because of their covariate nature.

Types of insurance

Hopes some years ago that the smallholder insurance industry might take off have been disappointed (Hazell, 1992). However, efforts to develop appropriate insurance products for the covariate risks of developing country agriculture have continued – such as risk securitization in Nicaragua (Miranda and Vedenov, 2001). Over the last few years, there has been an increasing interest in hedging against weather-related events, following the increasing likelihood of extreme phenomena related to global environmental issues such as climate change. Public intervention in this area is now growing, with governments looking with interest towards market-based tools to hedge financial positions in case of adverse events. Newer and promising products, at both micro and macro levels, are indexed insurances. These are based on indemnities tied to predetermined indicators for a common geographical area, unlike the assessment of losses sustained by individual enterprises which are examined in the field. Indirect-index insurance, also called weather derivatives, falls in this category (Schaffnit-Chatterjee, 2010: 23–4). Mahul and Stutley (2010) cite China, India, and Mexico as countries where provision of weather insurance has expanded. Differences between indemnity and index-based agricultural insurance products are summarized in Table 4.1.

At the micro level also, insurance is considered to be a promising area for promoting agriculture in poor developing countries, as well as for managing safety nets and disaster management, as shown by the increasing number of projects, pilots, experiments, and even policy schemes which are being undertaken that entail some sort of agricultural insurance (Mahul and

Table 4.1 Differences between indemnity and index-based agricultural insurance

	Indemnity-based insurance products	Index-based insurance products
	Losses assessed at individual level	Losses assessed using an index
Crop insurance products	Damage-based products – frost, hail, fire, floods, storms, pests, diseases ... Theft-based products Yield-based products Shortfall and crop revenue insurance	Area yield-based index insurance Weather index-based insurance Normalized difference vegetation index insurance
Livestock insurance products	Accident and mortality insurance Herd insurance Epidemic disease insurance	Mortality risk insurance

Source: adapted from: Mahul and Stutley (2010)

Stutley, 2010). This market potential is increasingly recognized by insurance providers in advanced economies:

> The success of microcredit worldwide has shown that people with low incomes are a proven market for financial services and are effective consumers if given appropriate products, processes, and knowledge. In the insurance field, microinsurance can provide the specialised insurance products demanded by under-served low income markets (Lloyd's Micro Insurance Centre, undated: 3).

Nevertheless, there are complex problems concerning agricultural insurance in developing countries:

> Small producers in developing countries want it mainly to cover loss of crops, livestock, plantations and farm equipment. Challenges for insurers include the high cost of distribution, the high costs of loss assessment and claims handling, and difficulty in controlling fraud and moral hazard. Local schemes struggle with correlated risks, such as droughts, that affect whole regions and therefore most clients (Lloyd's Micro Insurance Centre, undated: 26).

While market demand opportunities are becoming clearer, the specialist nature of provision is also evident, as shown in Table 4.2.

Resuming investment to enhance productivity and production in poor countries requires a number of combined actions, and the reduction of exposure to risks can play an important role. Financial schemes involving insurance are considered to be one way of supporting agriculture. Insurance does not destroy market incentives as some emergency or humanitarian financial or physical product interventions do, but rather builds on the potential for private suppliers to be involved in financial markets.

Government support can take various forms. In the first instance, insurance legislation is needed to create an enabling environment through insurance legislation. Secondly, direct interventions can include premium subsidies and reinsurance. Nevertheless, experiences of subsidies to private-public partnerships tend to be mixed in terms of financial sustainability and market development. Many schemes have been abandoned, and many markets have failed to develop due to lack of interest from farmers, who simply did not purchase the policies. Finally, training and information provision about risk management, monitoring, and early warning systems are simple and necessary forms of intervention that directly increase people's individual and organizational capacity to manage risks and livelihood strategies.

The Global Index Insurance Facility (GIIF) of the International Finance Corporation, working with private financial services providers such as Swiss Re and rural sector firms such as the Syngenta Foundation, is one such publicly sponsored scheme intended to provide insurance cover in developing countries. Cover is provided particularly to smallholder farmers who, on the one hand, experience problems of lack of information and expert financial and

Table 4.2 Differences between traditional insurance and microinsurance

	Traditional insurance	Microinsurance
Clients	Low risk environment Established insurance culture	Higher risk exposure/high vulnerability Weak insurance culture
Distribution models	Sold by licensed intermediaries or by insurance companies to wealthy clients or companies that understand insurance	Sold by non-traditional intermediaries to clients with little experience of insurance
Policies	Complex policy documents with many exclusions	Simple language Few exclusions Group policies
Premium calculation	Monthly to yearly payments, often paid by mail based on an invoice, or by debit orders	Frequent and irregular payments adapted to volatile cash flows of clients Linked with other transactions (e.g. loan repayment)
Control of insurance risk (adverse selection, moral hazard, fraud)	Limited eligibility Significant documentation required Screenings, such as medical tests, may be required	Broad eligibility Limited but effective controls (reduce costs) Insurance risk included in premiums rather than controlled by exclusions Link with other services (credit)
Claims handling	Complicated processes Extensive verification documentation	Simple and fast procedures for small sums Efficient fraud control

Source: adapted from: Lloyd's Micro Insurance Centre (undated: 7–8)

risk management knowledge, and on the other are subject to systemic hazards for which there is under-provision by the private sector. Using predefined statistical indices of events, such as the natural disasters to which agriculture is prone, to trigger payouts speeds response to disaster and reduces the costs of administration and individual loss adjustment (Box 4.3).

The role of governments in insurance provision

Very few experiences of agricultural insurance work on a purely market basis, and even several of the more innovative indexed policies require support from the public sector or foreign donors. This is the case in countries where agriculture enjoys extensive support, such as the US and the EU: such public supports tend to crowd out private insurance provision, so that subsidies are required to stimulate demand. Equally, where agricultural insurance appears in developing countries, public support is not uncommon. Especially where local resources are more constrained, foreign and/or donor resources are employed to finance agricultural insurance schemes, at least for facilitating reinsurance. For poor farmers in developing countries, motivations for

> **Box 4.3 Index-based insurance for rural communities**
>
> The Global Index Insurance Facility (GIIF) is a multi-donor trust fund that supports the development and growth of local markets for indexed/catastrophic insurance in developing countries, primarily in sub-Saharan Africa, Latin America and the Caribbean, and Asia Pacific. GIIF's objectives are to provide access to finance to vulnerable populations and to expand the use of index insurance as a risk management tool in agriculture, food security, and disaster risk reduction. GIIF's implementing partners covered more than 1.3 million farmers, pastoralists, and microentrepreneurs as of July 2016, with $148 million in sums insured.
>
> Providing access to finance for the vulnerable, insurance is an important element of poverty alleviation. Unfortunately, agricultural insurance and disaster insurance are either unavailable or prohibitively expensive in many developing countries. Index insurance is an innovative approach to insurance provision that pays out benefits on the basis of a predetermined index or loss of assets and investments resulting from weather and catastrophic events, without requiring the traditional services of insurance claims assessors. It also allows for the claims settlement process to be quicker and more objective.
>
> *Source*: International Finance Corporation (2016)

public intervention in agricultural insurance range from the need to reinsure systemic risks, to the need to overcome start-up information and transaction costs, and because premiums may be unaffordable for poor households.

In general, governments seem to intervene well beyond what they would be expected to do: subsidization of premiums and financial support to private insurance companies or parastatals is widespread and goes beyond the correction of market failures generated by catastrophic risks and regulation and the provision of information. While this can benefit insurance companies, the effect on agriculture is not always clear. In fact, premium subsidization is often associated with reports of inefficiency and fraud. Targeting insurance towards poorer farmers is difficult and many schemes fail to address the needs of the poorest.

Because insurance is one element contributing to the removal of obstacles to agricultural investment and the development of smallholder farming, policies should be consistent with the broader objective of promoting a complete and functioning market for all agricultural financial services. In fact, insurance has frequently developed in connection with the market for other services, particularly credit: in summary, access to credit and insurance are mutually reinforcing.

Economic risk and types of value chains

Having surveyed the issues concerning producers' price-risk management, we turn to value chains in developing country contexts. Factors affecting risk management in different value chains will be driven essentially by product characteristics such as:

- the degree of product perishability, which dictates the feasibility of storage and long-distance transportation;

- the degree of product quality standardization or differentiation, which may either be determined by the product's nature and geographic origin, or by marketing promotion policies of supply-chain managers, and which greatly determine the degree of market power that producers may achieve;
- the degree of processing, or the extent to which the bulk product is used also as an input for other industries, which may condition the characteristics above.

For example, using the agricultural product as an input in a processing activity may increase the storability of the products and drive a price and processing 'wedge' between consumers and producers of the bulk commodity. We have noted that the consumer markets for beverages products such as coffee or chocolate are distinct and remote from the producer markets for raw coffee or cocoa in which smallholders are massively involved. If the bulk agricultural product is used as an input in other industries, the price variation for the agricultural product is only indirectly linked to the evolution of final demand. Risks can be seen to be a property of business throughout the value chain.

Figure 4.3 summarizes the types of risks that are common to agri-food chains, occurring at different stages: risks external to the chain but which affect farm and firm performance are policy, supply, and distributional issues; risks internal to the chain include technology, production, information, and organizational risks.

International fresh produce-chains

The capacity to supply international markets with fresh, perishable, unprocessed, speciality food products offers greater value-adding opportunities to chain participants, including those farmers who can overcome entry barriers of scale, management, and specification. However, this poses questions in terms of the kind of risks to which the chain is exposed and of the risk management options available for producers. These are related to the perennial nature of some crops – such as tree crops – and therefore the inelastic supply response which can create price instability when demand fluctuates. For perennial and other crops there are also questions about quantities, delivery, and quality assurance. And of major importance in international markets are trade risks due to political instability and price risks due to fluctuations in exchange rates.

Depending on product characteristics, the chain organization will differ along the following dimensions, which are relevant for the objective of exploring the impact on producers of the basic agricultural product, especially small farmers:

- level of complexity/integration;
- extent of control operated by the various agents;
- locus of price formation;
- major production risks, other than price variation, to which the producer is exposed.

RISK MANAGEMENT FOR AGRICULTURAL SMALLHOLDERS 103

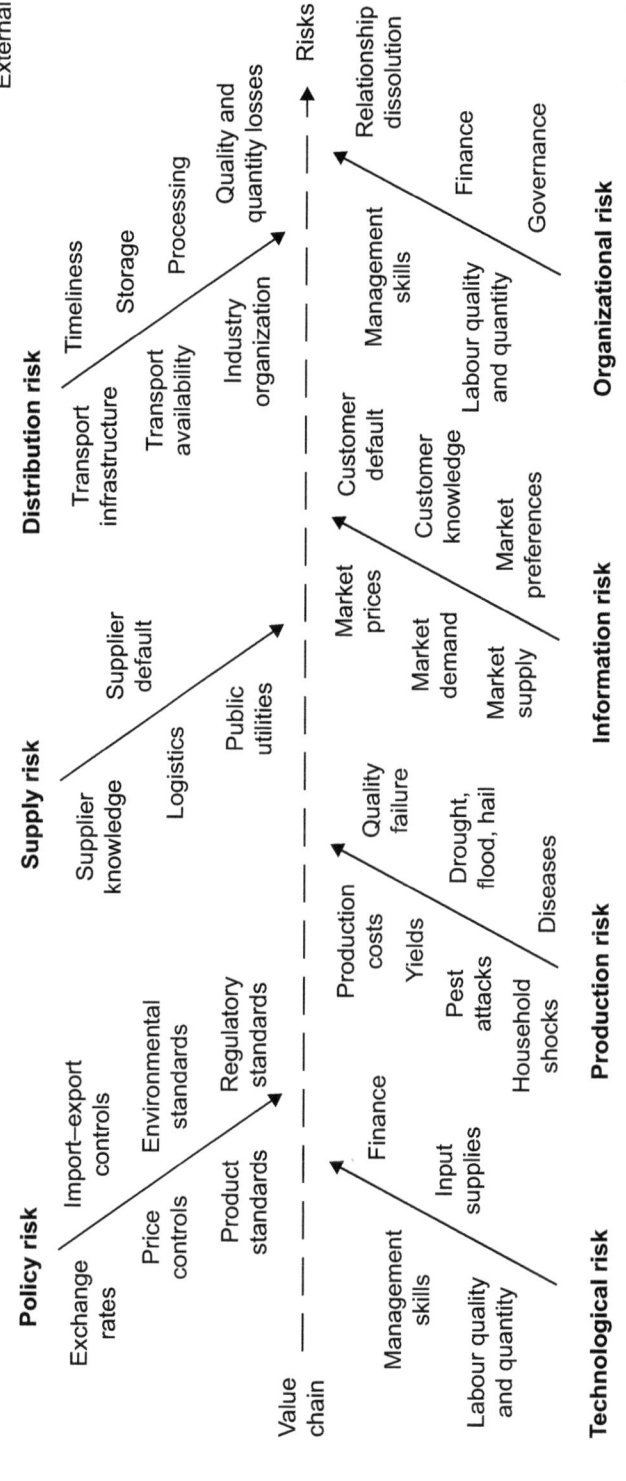

Figure 4.3 Value chain risk management
Source: adapted from Lin and Zhou (2011)

These, in turn, will determine:

- likelihood that small farmers will participate in the value chain;
- share of value added that can be appropriated by small farmers;
- potential for income-risk reduction for farmers by participating in the value chain;
- role for effective public action.

Staple foods for domestic consumption

The production of staple foods is still the most important sector in many agriculture-based and emerging economy countries (World Bank, 2007: 118). There are two major destinations for staple foods: domestic consumption and export. The share of total production that is destined for domestic consumption still forms a large share of agricultural production, especially the output of small farmers.

Value chain organization in these cases can vary in complexity. The least complex are short chains from producers, through one or two or no intermediaries, to local consumers in the village. Increasingly complex chains involve multiple traders who concentrate supply in the producing areas, serve wholesale markets, include sophisticated market functions of storage, transport, and processing, and distribute to urban retailers and export markets.

Often there is some involvement of government in procurement, storage, and distribution activities, particularly of staple crops: examples can be found in the Public Food Distribution System (PDS) in India (Pritchard et al., 2014), or the Food Reserve Agency (FRA) (Poole, 2010) in Zambia. The level of formal or informal contractual agreements of farmers in these cases is limited. Governments have often tried to facilitate more complex and risk-limiting contractual arrangements in developing countries through the development of commodity exchanges, with mixed success (Jayne et al., 2014; Meijerink et al., 2014). For example, in Zambia, Sitko and Jayne (2012) point to a series of factors that restrict the development of the market: an unwillingness of financial institutions to engage, asymmetric information among traders leading to opportunistic behaviour and therefore lack of trust, conflicts of interest, market manipulation, and high costs of trading. Finally, unpredictable interventions of government in the cereals market impede development (Sitko and Jayne, 2012).

The spread of warehouse receipt systems, usually for staple foods, is another example of market procurement facilitated by government support that addresses the private objectives of individual farmers and the public objective of national food security (Box 4.4).

Where the major market is domestic consumption, the price formation locus is the retail market, where the final purchase occurs. It is determined partly by the local supply and demand conditions, and sometimes by government intervention, usually for consumer protection. This may involve

> **Box 4.4 Risk management through aligning private and public objectives**
>
> 'Warehouse receipt systems allow agricultural producers to access credit by borrowing against receipts issued for goods stored in independently controlled warehouses. These systems enable producers to delay the sale of their products until after harvest, to a moment when prices are generally more favourable. Warehouse receipt systems can therefore mobilize credit for the agricultural sector and improve agricultural trade.
>
> 'Warehouse receipt systems bring other benefits to the agricultural sector. For example, the provision of good handling and storage by warehouses licensed and controlled according to mandatory standards can help reduce post-harvest losses and improve product quality. The increased storage of agricultural commodities after the harvest season may in turn contribute to stabilizing commodity price volatility. With respect to the management of national food security and strategic reserves, an effective warehouse system provides government authorities with timely and accurate information about the aggregate stock of stored agricultural commodities in the country ...
>
> 'In recent years, many countries around the world have begun to introduce or reform legislation of their warehouse receipt system. Various objectives motivated the reforms, from mobilizing credit for the agricultural sector after market liberalization to adapting an existing system to the requirements of electronic commerce, or enhancing the income of smallholder producers.'
>
> *Source*: Wehling and Garthwaite (2015: vi)

various forms of price control and stabilization initiatives considered to be easily implementable, which often involve public storage.

Although price risk should not be of great concern to producers who enter into contracts with the traders or the government establishments, there is arguably scope for improvement of the market-information systems that link producers to the final market. This can be of great benefit to producers in remote areas, who might be excluded from the expanding consumer markets by post-harvest losses due to poor storage infrastructure, wide marketing margins that reduce returns, poor market integration, limited access to finance, and weak regulatory institutions. Under such conditions, improvements in the marketing infrastructure and in the market-information system can enable farmers to enhance the contractual terms obtained from traders.

Apart from price risk, yield risks and personal accidents are likely to be the most relevant risks which reduce the income derived from staple foods farming, and compromise the ability of farmers to repay debts.

Fresh fruits and vegetables for domestic consumption

In more developed fresh fruits and vegetables chains, the commercial fruit and vegetable producers are usually not among the asset-poorest farmers: they will have irrigation, make intensive use of labour, and have already been able to step up from subsistence and staple food production (Dorward, 2009). On the other hand, for fresh fruits and vegetables, the possibility of arbitraging through storage is precluded because of the perishability of the products.

As reported above, traders may have to procure and pay for the product through advance contracts, thus bearing most of the risk associated with product deterioration and other risks, which may be particularly high in many developing countries owing to poor infrastructure.

Here too, the supply chain risk is particularly vulnerable to natural shocks. Weather effects are less likely to impact more sophisticated production systems that are often physically protected from environmental impacts, and irrigated. Environmental shocks are more likely to affect post-harvest bulking and distribution activities; for example, through weather-related damage to roads and bridges. Investment in public infrastructure and the provision of medium-term loans to private traders are likely to be very effective means to reduce the overall risk exposure in the chain and to thus benefit the farmer suppliers.

Although open-air markets and roadside kiosks are still the dominant forms of retail for fresh fruits and vegetables, rapid change in retail organization is progressing in almost all developing regions. Supermarkets are slower to gain market dominance in the sale of fresh fruits and vegetables in developing countries, but the modernization of other parts of the chain will likely affect the organization of the supply chains for fresh produce (Cadilhon et al., 2006; Reardon et al., 2009; Minten et al., 2010). If nothing else, there will be a segmentation of the market through sizeable differences in prices among vegetables and fruits sold in the supermarkets and those sold in traditional open-air markets and roadside kiosks. A growing share of the production will be procured by large retailers and better-organized traders, through production contracts rather than through open market transactions, when timing and quality of production becomes crucial.

Internationally traded commodities

For commodities such as coffee, cocoa, tea, and cotton for which there is an active international trade, the central locus where the price is formed is usually abroad, and the scope for influencing the process of formation of such prices by producers, consumers, or governments in developing countries is very limited (Borzoni and Poole, 2011).

If the market were free from distortions, producers would be exposed to the price risk associated with the variability of international markets. In most commodity markets, however, governments can, and do, influence the effective price that producers receive for their product by various forms of trade policies, thus breaking the link between internationally quoted prices and effective farm-gate prices. The issue of price risk for producers becomes one of institutional risk; that is, the risk that the government might change policy in unexpected and adverse ways.

Although producers themselves might not be able to use futures and options to hedge price risk, the international dimension of the market and the presence of international commodity exchanges for these products nevertheless provide important scope for intermediaries to hedge against negative price variations, and they may then transfer part of the benefits to farmers in terms of more stable prices.

One other change that is occurring in the supply chains of traditional bulk commodities such as coffee, cocoa, and tea, is the attempt to colonize new market niches, resulting in a wave of diversification into speciality commodities, up to the point of causing the switch to a distinctly different commodity chain organization.

Speciality commodities for international markets

One significant evolution in the markets for traditional food products such as coffee, chocolate, and tea has been the creation of segmented niches based on fair trade or environmental sustainability characteristics to try to attract rich consumers. This has been done by attaching to the product some 'quality' dimension in an attempt to diversify buyers away from the mature, traditional, bulky markets. Brands, denominations of origin, organic certified, fair trade, bird-friendly, socially responsible, etc. are very common labels attached to the more-less traditional products to extract a greater value-added premium from the consumers (Donovan and Poole, 2014a, b).

The fundamental reasons for the development of this type of institution and/or participation by large private companies involved in the trade, processing, and retail of agricultural commodities can be questioned: is it really a genuine interest in the conditions of the poor, a superficial acknowledgement of corporate social responsibility, or is it just seen as an opportunity to colonize and revitalize old markets through smart and innovative product differentiation? Ensuring stable access to remunerative export markets is a major benefit to producers (Donovan and Poole, 2014b). Nevertheless, research in Ethiopia and Uganda found that fair-trade certified coffee and tea arguably do not improve lives of the very poorest (Cramer et al., 2016).

Switching away from a traditional, undifferentiated product towards an added-value product has implications for the characteristics of the demand faced by retailers. On one hand, the price elasticity of demand is dramatically reduced, which allows sellers to differentiate their product from the competing ones and to sustain price mark-ups over marginal costs; on the other hand, the demand for these products tends to be more volatile and extremely sensitive to perceived drops in the reliability of the quality information and to changing economic conditions (García Martínez and Poole, 2008).

Product differentiation is a fundamental strategy for marketers of food products. However, most of the attributes that are more relevant are of a 'credence' type: the truthfulness and the actual content of labels such as GMO free, organic, pesticide free, fair trade, socially responsible, and denomination of origin cannot be objectively assessed by the consumers, who must thus rely on some form of information which can be either privately or publicly provided. When the certification is reliable and where strong institutional settings exist to enforce the truthfulness of labelling, retailers have to coordinate and control product management across the entire supply chain. This will imply a stronger integration with producers, which could be obtained through complete vertical integration. Forms of contractual relationship between producers and other agents can definitely allow producers, even in developing countries, to participate in the distribution of the added value generated by product diversification. They will usually require investment by the producers to comply with special production and a traceability system which allows for producer identification.

Associated with these requirements, new types of risk for producers arise. First, after investments are realized, the other participants in the chain may fail to accept delivery of the product if a cheaper source of production becomes available, or if demand conditions change. Thus, 'agreements' may not be binding and buyers can default on purchase commitments. Second, 'sunk' or irrecoverable investments in specific assets may weaken producers' position and induce them to accept a lower share of the value added. Third, there is risk of some type of failure in the value chain that compromises product quality, with producers held liable.

These risk considerations may condition the participation of developing country producers in highly coordinated value chains. For producers to profitably participate in these chain opportunities depends in part on an effective institutional and legal system for contract design and enforcement.

Fresh fruits, vegetables, and flowers for international markets

Another rapidly growing sector in the global agricultural trade from the South to the North is the one that involves highly perishable products such as fish, meat, fresh fruits and vegetables, and cut flowers. Such products account for about 50 per cent of the value of agricultural exports from developing countries. The production, marketing, and trade of fresh and processed fruits and vegetables has been described as 'one of the most dynamic segments of developing country participation in world markets' (Minten et al., 2009: 1), making global chains both accessible and lucrative for poor farmers.

The problems and pressures on supply chain integration posed by these products are similar to those analysed for speciality foods. The value added is linked to the ability of sellers to diversify products by associating with the product some quality characteristics for which rich consumers are willing to pay. However, in these cases the quality attributes are of a 'search' type, more easily assessed by consumers before purchase, and therefore their supply is mostly affected by timing of the delivery, compliance with detectable sanitary and phytosanitary standards, variety of choice, and freshness, all things which producers, above anyone else, control (Poole et al., 2007). Fair-trade chains are found in such markets, and the research in Uganda and Ethiopia referred to earlier (Cramer et al., 2016) found against fair trade in the cut-flower market also.

For some standard fruits, such as bananas and pineapples, the dominant configuration in global trade in the past has been vertical integration, with trading companies taking direct control of production through ownership of local plantations. In that context, risk is (almost) fully internalized, and the role of small farmers is non-existent, other than as wage workers in the plantations, often in very weak bargaining positions.

Nevertheless, the spread of information on the conditions of workers in the plantations and corporate social responsibility has impelled change even in the chains of these standardized tropical fruits, with more attention given to socially responsible practices and certification. At the same time, for other fresh products which need faster and more careful handling, the prevailing model is contractual agreements between independent producers and traders or retailers. This offers producers a stronger contractual position within the value chain, although high asset thresholds for individual smallholder participation remain a barrier. Collective organization tends to be a prerequisite for participation, improved risk management, and investment in transport, handling, and packaging infrastructure. Box 4.5 gives a brief account of a case which illustrates issues of contractual arrangements and risk management between a large export firm and small producers.

The final section of the book will explore other cases where projects have aimed to integrate producers into commercial markets.

Box 4.5 French beans from Madagascar

French beans in Madagascar are procured and sold to supermarkets in Europe by one Malagasy company, Lecofruit SA. The results of a survey of 200 representative farmers producing vegetables under contracts in the Highlands of Madagascar showed that contracting was beneficial for farmers, reducing the lean inter-seasonal period for producers, diversifying income sources and thus providing income insurance.

The contract characteristics allowed interesting observations related to the risk implications for the farmers and for Lecofruit SA to be made:

- Farmers retained most of the production risk, with the additional dimension of quality risk: the contract specified that the firm would only pay for product that fulfils the quality norms set in advance.
- Because it is in the interest of the firm to maintain adequate supply, Lecofruit SA provided technical assistance and training, in addition to monitoring and supervision of the contracted areas.
- Potential alternative market outlets for producers created incentives for Lecofruit SA to offer a price premium to attract and retain farmers and to ensure enforcement.
- Contracts proved to be self-enforcing, necessarily so given the poorly developed legal institution and the relatively high transaction costs for formal enforcement, compared with the small amount involved in any individual contract.
- Contracts appeared to be robust even under conditions of high inflation and therefore price risk.

Source: Minten et al. (2009)

Conclusions

The fast-growing interest that has formed around the concept of value chains in developing-country agriculture has revealed that there are important opportunities for income growth and expanding markets that can be captured by farmers in these countries. At the same time, engagement in such markets increases the level of risk faced by small-scale farmers. The major lessons that can be derived from the discussion of the many problems related to increasing risk exposure and risk management in the previous pages are outlined below.

Exposure to uninsured risk, in general, still determines welfare losses to poor people in developing countries. Risk is still one of the major determinants of the agricultural production and food value chains of most developing countries, and ultimately of the real opportunities available to small-scale farmers.

Getting involved in value chains for new and evolving markets for agri-food products does not necessarily reduce the risk exposure of those who have less bargaining power within the chain. Product characteristics, such as storability and quality standardization, can reduce the ability of farmers to negotiate formal and informal contractual agreements that dominate the value chain transactions, but can provide some safeguards in equilibrating negotiating power. In value chains for products such as grains, or bulk commodities such as coffee, tea, and sugar, producers remain price takers, exposed to price risk beyond their own marketing decisions. The ability to manage income risk depends on access to credit, insurance, and contract terms. Public policy is needed to strengthen market-supporting institutions to improve access to financial services and mechanisms for contract enforcement. Fostering collective action creates advantages in accessing markets and handling risk, as well as raising important governance challenges.

For products with a low degree of storability, such as fresh fruits and vegetables, and fish and meat products, farmers will have stronger bargaining power, especially where product quality and timing of delivery is a crucial element in attracting high willingness-to-pay consumers, both in domestic urban areas and in export markets. In the value chain organization, farmers can retain most of the risk related to variation of quality of their product, despite technical assistance provided by traders and retailers. Also, traceability and other labelling requirements impose on producers some costs and risks that are not present in traditional marketing systems. An important risk management option in these cases is linked to investments in education and technical innovation. Government participation remains crucial in assisting agricultural producers in developing countries to manage their risk, although public interventions should be limited to investment in infrastructure and institutions to promote three conditions:

- information sharing;
- competition;
- contract enforcement.

Indeed, the prevailing view of the role of the state goes beyond these potential roles to create an environment in which agribusiness can operate fairly and efficiently, all the time balancing the need for specific interventions with the need to create a climate that enables farmers and firms to unleash their own business initiatives:

> Governments can help by establishing appropriate regulatory systems that ensure the safety and quality of agricultural goods and services without being costly or burdensome overall so as to discourage firms from entering the market. Excessive regulation makes firms move to the informal economy and generates high unemployment. Poorly-designed regulations impose high transaction costs on firms thus reducing trade volumes, productivity and access to finance. Creating an enabling environment for agriculture is a prerequisite to unleash the sector's potential to boost growth, reduce poverty and inequality, provide food security and deliver environmental services. Among other factors, government policies and regulations play a key role in shaping the business environment through their impacts on costs, risks and barriers to competition for various players in the value chains. By setting the right institutional and regulatory framework, governments can help increase the competitiveness of farmers and agricultural entrepreneurs, enabling them to integrate into regional and global markets (World Bank, 2017: x).

References

Barrett, C.B. (2008). Smallholder market participation: concepts and evidence from eastern and southern Africa. *Food Policy* **33**(4): 299–317 <https://doi.org/10.1016/j.foodpol.2007.10.005>.

Borzoni, M. and Poole, N.D. (2011). Sourcing strategies in the Italian coffee industry. *Food Chain* **1**(1): 71–86 <http://dx.doi.org/10.3362/2046-1887.2011.006>.

Cadilhon, J.-J., Moustier, P., Poole, N.D., Giac Tam, P.T. and Fearne, A. (2006). Traditional vs. modern food systems? Insights from vegetable supply chains to Ho Chi Minh City (Vietnam). *Development Policy Review* **24**(1): 31–49 <http://dx.doi.org/10.1111/j.1467-7679.2006.00312.x>.

Cafiero, C. (2008). *Agricultural Producer Risk Management in a Value Chain Context: Implications for Developing Countries' Agriculture.* EU-AAACP Paper Series. No. 4. Rome, Food and Agriculture Organization of the United Nations. Retrieved 28 March 2017, from http://www.fao.org/fileadmin/templates/est/AAACP/inter-regional/FAO_AAACP_Paper_Series_No_4_1_.pdf.

Conforti, P. (2009). *Agricultural Insurances as Market Based Tools for Risk Management and Agricultural Development.* EU-AAACP Paper Series. No. 5. Rome, Food and Agriculture Organization of the United Nations. Retrieved 28 March 2017, from http://www.fao.org/fileadmin/templates/est/AAACP/inter-regional/FAO_AAACP_Paper_Series_No_5_1_.pdf

Cramer, C., Johnston, D., Mueller, B., Oya, C. and Sender, J. (2016). Fairtrade and labour markets in Ethiopia and Uganda. *Journal of Development Studies*: 1–16 <http://dx.doi.org/10.1080/00220388.2016.1208175>.

Dercon, S., ed. (2005). *Insurance Against Poverty*. Oxford, Oxford Scholarship Online.

Donovan, J. and Poole, N.D. (2014a). Partnerships in fairtrade coffee: a close-up look at how buyers and NGOs build supply capacity in Nicaragua. *Food Chain* **4**(1): 34–48 <http://dx.doi.org/10.3362/2046-1887.2014.004>.

Donovan, J. and Poole, N.D. (2014b). Changing asset endowments and smallholder participation in higher value markets: evidence from certified coffee producers in Nicaragua. *Food Policy* **44**: 1–13 <http://dx.doi.org/10.1016/j.foodpol.2013.09.010>.

Dorward, A. (2009). Integrating contested aspirations, processes and policy: development as hanging in, stepping up and stepping out. *Development Policy Review* **27**(2): 131–146 <http://dx.doi.org/10.1111/j.1467-7679.2009.00439.x>.

Fafchamps, M. and Minten, B. (1999). Relationships and traders in Madagascar. *Journal of Development Studies* **35**(6): 1–35 <http://dx.doi.org/10.1080/00220389908422600>.

FEWS NET (undated). Home page. Retrieved 9 May 2017, from http://www.fews.net/

García Martínez, M. and Poole, N.D. (2008). *The development of private fresh produce safety standards: Implications for developing Mediterranean exporting countries. Fresh Perspectives 26. agrifoodstandards.net*. Retrieved 28 March 2017, from https://assets.publishing.service.gov.uk/media/57a08b96e5274a31e0000c4a/60506-fp26.pdf.

Hazell, P.B.R. (1992). The appropriate role of agricultural insurance in developing countries. *Journal of International Development* **4**(6): 567–581 <http://dx.doi.org/10.1002/jid.3380040602>.

Hazell, P., Pomareda, C. and Valdés, A., eds (1986). *Crop Insurance for Agricultural Development: Issues and experience*. Baltimore, MD, Johns Hopkins University Press.

International Finance Corporation (2016). Global Index Insurance Facility. Retrieved 28 March 2017, from http://www.ifc.org/wps/wcm/connect/industry_ext_content/ifc_external_corporate_site/financial+institutions/priorities/access_essential+financial+services/global+index+insurance+facility.

Jaffee, S., Siegel, P. and Andrews, C. (2008). *Rapid Agricultural Supply Chain Risk Assessment: Conceptual framework and guidelines for application*. Commodity Risk Management Group, Agricultural and Rural Development Department. Washington, DC, World Bank. Retrieved 28 March 2017, from http://siteresources.worldbank.org/INTCOMRISMAN/Resources/RapidAgriculturalSupplyChainRiskAssessmentConceptualFramework.pdf.

Jayne, T.S., Sturgess, C., Kopicki, R. and Sitko, N. (2014). *Agricultural Commodity Exchanges and the Development of Grain Markets and Trade in Africa: A review of recent experience*. Working Paper 88. Lusaka, Zambia, Indaba Agricultural Policy Research Institute (IAPRI). Retrieved 28 March 2017, from http://ageconsearch.umn.edu/bitstream/188568/2/wp88.pdf.

Johnston, D. and Morduch, J. (2008). The unbanked: evidence from Indonesia. *World Bank Research Observer* **22**(3): 517–537 <http://dx.doi.org/10.1093/wber/lhn016>.

Lin, Y. and Zhou, L. (2011). The impacts of product design changes on supply chain risk: a case study. *International Journal of Physical Distribution & Logistics Management* **41**(2): 162–186 <http://dx.doi.org/10.1108/09600031111118549>.

Lloyd's Micro Insurance Centre (undated). *Insurance in Developing Countries: Exploring Opportunities in Microinsurance*. Lloyds 360° Risk Insight. Retrieved 28 March 2017, from https://www.lloyds.com/news-and-insight/risk-insight/library/society-and-security/insurance-in-developing-countries.

Mahul, O. and Stutley, C.J. (2010). *Government Support to Agricultural Insurance : Challenges and options for developing countries*. Washington, DC, World Bank. Retrieved 9 May 2017, from https://openknowledge.worldbank.org/handle/10986/2432.

Meijerink, G., Bulte, E. and Alemu, D. (2014). Formal institutions and social capital in value chains: the case of the Ethiopian Commodity Exchange. *Food Policy* **49**(1): 1–12 <https://doi.org/10.1016/j.foodpol.2014.05.015>.

Minten, B., Randrianarison, L. and Swinnen, J.F.M. (2009). Global retail chains and poor farmers: evidence from Madagascar. *World Development* **37**(11): 1728–1741 <https://doi.org/10.1016/j.worlddev.2008.08.024>.

Minten, B., Reardon, T. and Sutradhar, R. (2010). Food prices and modern retail: the case of Delhi. *World Development* **38**(12): 1775–1787 <https://doi.org/10.1016/j.worlddev.2010.04.002>.

Miranda, M. and Vedenov, D.V. (2001). Innovations in agricultural and natural disaster insurance. *American Journal of Agricultural Economics* **83**(3): 650–655 <https://doi.org/10.1111/0002-9092.00185>.

North, D.C. (1987). Institutions, transaction costs and economic growth. *Economic Enquiry* **25**(3): 419–428 <https://doi.org/10.1111/j.1465-7295.1987.tb00750.x>.

North, D.C. (1990). *Institutions, Institutional Change and Economic Performance*. Cambridge, Cambridge University Press.

North, D.C. (1994). Economic performance through time. *American Economic Review* **84**(3): 359–368.

Poole, N.D. (2000). Production and marketing strategies of Spanish citrus farmers. *Journal of Agricultural Economics* **51**(2): 210–223 <https://doi.org/10.1111/j.1477-9552.2000.tb01224.x>.

Poole, N.D. (2010). *Zambia Cassava Sector Policy: Recommendations in support of strategy implementation*. EU-AAACP Paper Series. No. 16. Rome, Food and Agriculture Organization of the United Nations. Retrieved 28 March 2017, from http://www.fao.org/fileadmin/templates/est/AAACP/pacific/07_FAO_AAACP_Paper_Series16_Recommendations_Zambia_Cassava_Strat.pdf.

Poole, N.D., Del Campo Gomis, F.J., Juliá Igual, J.F. and Vidal Giménez, F. (1998). Formal contracts in fresh produce markets. *Food Policy* **23**(2): 131–142 <https://doi.org/10.1016/S0306-9192(98)00024-4>.

Poole, N.D., Seini, A.W. and Heh, V. (2003). Improving agrifood marketing in developing economies: contracts in Ghanaian vegetable markets. *Development in Practice* **13**(5): 551–557 <http://dx.doi.org/10.1080/0961452032000166483>.

Poole, N.D., Martínez-Carrasco Martínez, L. and Vidal Giménez, F. (2007). Quality perceptions under evolving information conditions: implications for diet, health and consumer satisfaction. *Food Policy* **32**(2): 175–188 <https://doi.org/10.1016/j.foodpol.2006.05.004>.

Pritchard, B., Rammohan, A., Sekher, M., Parasuraman, S. and Choitani, C. (2014). *Feeding India: Livelihoods, entitlement and capabilities.* Abingdon, UK, Routledge.

Reardon, T., Barrett, C.B., Berdegué, J.A. and Swinnen, J.F.M. (2009). Agrifood industry transformation and small farmers in developing countries. *World Development* **37**(11): 1717–1727 <https://doi.org/10.1016/j.worlddev.2008.08.023>.

Schaffnit-Chatterjee, C. (2010). *Risk Management in Agriculture: Towards Market Solutions in the EU.* Frankfurt am Main, Deutsche Bank Research. Retrieved 11 November 2016, from http://www.dbresearch.com/PROD/DBR_INTERNET_EN-PROD/PROD0000000000262553.PDF.

Singh, I., Squire, L. and Strauss, J., eds (1986). *Agricultural Household Models: Extensions, applications, and policy.* Baltimore, MD, Johns Hopkins University Press.

Sitko, N.J. and Jayne, T.S. (2012). Why are African commodity exchanges languishing? A case study of the Zambian Agricultural Commodity Exchange. *Food Policy* **37**(3): 275–282 <https://doi.org/10.1016/j.foodpol.2012.02.015>.

Smith, A. (1776). *The Wealth of Nations* (Penguin Classics edition 1970). Harmondsworth, UK, Penguin.

Swenson, L. (2000). Income risk management: the perspective of United States farmers, in *Income Risk Management in Agriculture*, pp. 65–7. Paris, OECD. Retrieved 11 November 2016, from http://www.oecd.org/agriculture/agricultural-policies/42750750.pdf.

Wehling, P. and Garthwaite, B. (2015). *Designing Warehouse Receipt Legislation: Regulatory options and recent trends.* Rome, UN Food and Agriculture Organization. Retrieved 10 November 2016, from http://www.fao.org/3/a-i4318e.pdf.

World Bank (2007). *World Development Report 2008: Agriculture for Development.* Washington, DC, World Bank. Retrieved 28 March 2017, from http://siteresources.worldbank.org/INTWDR2008/Resources/WDR_00_book.pdf.

World Bank (2017). *Enabling the Business of Agriculture 2017.* Washington, DC, World Bank. Retrieved 28 March 2017, from http://eba.worldbank.org/~/media/WBG/AgriBusiness/Documents/Reports/2017/EBA2017-Report17.pdf.

PART II
Case studies of smallholder agriculture

CHAPTER 5

Introduction to Part II: Assessing the impact of commodity development projects on smallholder participation in agricultural markets: Case studies from Ethiopia, Peru, Tanzania, and Zambia

In this short chapter we prepare the way for the analyses that follow in Chapters 6–9 of market development projects in four developing countries, and the comparative analysis and synthesis in Chapter 10. The framework and methodology for the qualitative study are explained and limitations acknowledged. Conclusions are presented for readers who want to only skim Chapters 6–10.

Keywords: framework, assets, indicators, interventions, theory of change

Introduction

The principal theme of this book is to understand the constraints affecting smallholder farmer participation in agri-food markets and point to ways to help farmers take better advantage of market opportunities. In the earlier chapters we considered some of the constraints in theory and in practice. In the last chapter we considered risk and how to manage it. In Part 2 we report on four interventions, in different contexts, which aimed to increase farmers' commercial opportunities, and present simple analyses of the impacts of the projects. Risk is one of the factors that is highlighted. Chapters 6–10 draw substantially on papers by Amrouk et al. (2013) and Mudungwe et al. (2012).

Market-participation projects can be considered from two angles: first, and commonly, projects are designed by third-party organizations such as government bodies, international organizations, and non-governmental organizations. Whether or not these projects are designed and implemented in a participative manner, they can be considered as interventions by organizations that are inherently outside the agri-food value chain, as defined at the end of Chapter 1:

> *Intervention*: a project, action, or activity of an agency (public, private, third sector) external to the targeted beneficiaries as part of a policy formulated to achieve an objective.

Second, access to markets can come about through strategies and activities among participants within the value chain, and can be described better as initiatives, independent of outside interventions:

> *Initiative:* a project, action, or activity which arises from within or among beneficiary organizations and individuals.

Harper et al. (2015) argue that most business happens within commercial value chains that are not the object of project interventions. Their volume of case studies illustrates how farming smallholders and other poor people can profitably integrate their livelihood activities with modern value chains. And, they argue, projects are not needed because there are benefits for all value chain partners and it simply makes good business sense.

Nevertheless, we have seen that there are many reasons why smallholder farmers cannot easily access markets. The four interventions introduced here and reported in Chapters 6–9 were designed to overcome these difficulties. Four projects – on vegetables, milk, sisal, and henequen – were selected for analysis, and lessons about how these projects have succeeded in enabling better market participation by the smallholder farmers were drawn. Ultimately, the objective was to develop guidelines for best practices to be pursued in designing and implementing future commodity projects: 'to draw lessons on the determinants of successes and failures of projects assisting smallholder farmers to enhance their participation in agricultural value chains' (Amrouk et al., 2013).

Cases

The Food and Agriculture Organization of the United Nations (FAO) and the Common Fund for Commodities (CFC) have been implementing commodity-specific development projects for over 20 years in the context of the development plans agreed by the FAO Intergovernmental Groups (IGGs). Nearly 70 per cent of the projects have been located in least-developed countries. These projects mostly aimed to provide smallholder farmers with knowledge, skills, and services to increase agricultural productivity and household incomes. Project inputs to farming systems included dissemination of improved varieties, provision of fertilizers and pesticides, training on effective crop-management systems, and transmission of information on market trends and prices. In some instances, support to build market infrastructure, such as pack-houses for processing fruits and vegetables, greenhouses, and funding for the establishment of institutions such as farmers' cooperatives and associations, was also provided.

Since the scope to increase income through area expansion remains limited in many contexts, productivity growth through the adoption of new technologies and practices is required to enable smallholders to achieve higher returns. Higher productivity can therefore generate surpluses of marketable crops and livestock products, enabling better access to market opportunities for more consistent supplies of higher-quality and -quantity products.

However, the adoption of new technology for productivity gains is not straightforward. As we have noted in earlier chapters, in general, smallholders often cite lack of information about improved technologies, lack of access to credit, high risks associated with investments in agricultural technologies, or simply that the technology is not available, as the main constraints for commercializing agriculture. The success of any commodity-development project depends not only on a careful identification of the constraints, but also on the effectiveness of the proposed measures to overcome these constraints.

Four CFC/FAO-project case studies were selected for the analysis. These were identified following discussions with the secretaries of the FAO IGGs and officials at the CFC. These projects were chosen as they provided a good representative sample for the purpose of this study. In addition, other aspects were taken into account, including the availability of relevant documentation and national contacts, diversity in terms of commodity and geographical coverage, and logistical arrangements to minimize travel requirements for completion of the case studies.

The projects had different specific objectives. However, the overall goal of all the projects was to enhance the market-participation capacity of smallholders. Indicators based on livelihoods concepts were formulated to measure smallholders' market participation before and after the implementation of the project.

Most readers will want to move on to the following short chapters to gain an understanding of each intervention and read, mainly qualitative, summaries of the findings of and conclusions about the individual projects. For others who want a quick insight, we here give an overview of the four studies and make some overall comments on smallholder market participation.

A fuller account of each of the four cases is presented in Chapters 6–9. A quantitative summary and overview of lessons learned follows in Chapter 10.

Establishment of diversification programme for vegetable export development in Ethiopia and Sudan (CFC/FISGTF/17)

The purpose of the project was to strengthen the export capacity of smallholder vegetable farmers in Ethiopia and Sudan through the removal of critical supply-side constraints in relation to technical, infrastructural, business, and market factors. The project was implemented over a three-year period at an estimated cost of US$2 million. It had three focus areas:

- enhance the productive capacity of smallholders to export vegetable products, namely beans and okra;
- improve post-harvest handling skills and infrastructure;
- develop marketing and trading systems.

The immediate beneficiaries were farmers and out-growers. Institutions and organizations supporting farmers also benefited from training that was designed to enhance skills. The report which follows in Chapter 6 concerns only Ethiopia.

Production of oleaginous plants and commercialization of natural vegetable oils as substitutes for diesel fuel for public transportation in Peru and Honduras (CFC/FIGOOF/26)

The objective of this project was to promote the cultivation of rape and *Jatropha curcas* by smallholder farmers. The crops were to be processed and subsequently used as a substitute fuel by private and public commuter transport in cities in Peru and Honduras. The project commenced in April 2007 and was scheduled to be completed in March 2013. The estimated cost of the project was US$5.6 million. The direct project beneficiaries were smallholder farmers and small public transportation companies. The project consisted of the following five main components:

- development of production activities/techniques for rape and jatropha;
- production of vegetable oil by oil-extraction plants;
- substitution of diesel fuel by the extracted vegetable oil by private and public transportation companies;
- training activities delivered to various stakeholders (farmers, oil-production enterprises, and transport companies);
- dissemination of project results to different stakeholders.

The report which follows (Chapter 7) concerns only Peru.

Product and market development of Sisal and Henequen (CFC/FIGHF/07)

The objective of the project was to develop products and markets for sisal and henequen in Tanzania and Kenya. The project was implemented between January 1997 and December 2005. It had the following four components:

- development of new sisal varieties and improvement of cultivation practices;
- utilization of fibre-extraction waste for animal-feed production and for biogas and electricity generation;
- market studies and trials to establish the demand for sisal pulp and to identify potential buyers of the products;
- dissemination of project results.

The report in Chapter 8 concerns only Tanzania.

Strengthening the productivity and competitiveness of the smallholder dairy sector in Lesotho and Zambia (CFC/FIGMDP/14)

The project was implemented in Lesotho and Zambia over a four-year period at an estimated cost of US$3.3 million. It was aimed at improving the productivity and marketing position of smallholder dairy cooperatives.

The direct beneficiaries of the project were the smallholder dairy farmers, but also milk processors and consumers who benefited from increased local milk supplies and better-quality milk. Specifically, the project set out to achieve the following:

- promote better and more innovative livestock feeding technologies for local production and conservation of protein-rich feed stock;
- enhance milk quality and hygiene to reduce wastage and increase shelf life and safety of milk;
- pilot basic processing technologies to increase shelf life of milk targeted at different consumer groups and large-scale processors.

The report which follows in Chapter 9 concerns Zambia only.

Research approach and framework

Qualitative impact evaluation

Impact assessment is intended to determine whether a project had the desired effects on project beneficiaries, for example, farmers, households, and market institutions. Impact assessments can also explore unintended consequences, whether positive or negative, for beneficiaries. Some of the key questions are: How did the project affect the beneficiaries? Could programme design be modified to improve impact? Were the costs justified? In the cases which follow, the problem is enhancing market participation and the assessments were undertaken to answer some of these key questions.

Whereas a quantitative approach using randomization techniques is mainly suited for an *ex-ante* analysis, this was not appropriate for this *ex-post* analysis. Limited resources and time indicated the use of a non-experimental or quasi-experimental design to be appropriate. Qualitative impact evaluation was the principal approach used, and inferences were drawn from methods such as reviewing project implementation processes, interviewing project beneficiaries and other market participants, conducting focus-group discussions, and analysing supportive secondary data.

The CFC/FAO projects considered here faced many challenges due to the complex nature of the services delivered, the diversity of project activities, the different contextual environments, and the broad range of outcomes that such projects were designed to effect. Two evaluation approaches were used: theory-based evaluation, which is based on careful articulation of the programme model and use of the model as a guiding framework for evaluation; and participatory evaluation, which involved close collaboration between the evaluation team and project funders, supervisory bodies, executing agencies, and a sample of project beneficiaries. These two methods, used in combination, represent a powerful tool for meeting the ultimate objective of this task.

Theory-based evaluation involves identifying the key inputs and expected project outcomes, and analysing the underlying assumptions about how these inputs would lead to the desired outcomes. It implies the examining of the assumptions underlying the causal chain from inputs to outcomes and impact. The various links in the chain are analysed using a variety of methods, building up an argument as to whether the theory has been realized in practice. It traces how the (short-term) project activities and outputs will cause (short- to mid-term) outcomes and how these will lead in turn to (longer-term) social and economic impacts. This approach, as used here, is shown in Figure 5.1, which illustrates the pathway from the inputs made, in these cases, by external interventions which are expected to overcome market-entry barriers, not least by enriching the asset status of the beneficiaries. Leveraging these livelihood assets leads to a series of positive economic and other outcomes. Sustainability is achieved through the impact of immediate outcomes on the long-term enhancement of livelihood assets.

The development of quantitative indicators was guided by the causal linkages identified in Figure 5.1. Broadly, there were two groups of indicators: livelihood indicators of the participating farmers, and market indicators. The four case studies concerned commodities from four different groups, namely export vegetables, milk, sisal, and oleaginous plants. A set of specific indicators was adapted and analysed for each project.

Livelihood asset indicators

The first set of indicators related to the capacity or assets of smallholder farmers. A livelihood approach was used to assess change or how the projects have benefited the capacities of smallholder farmers. Livelihood

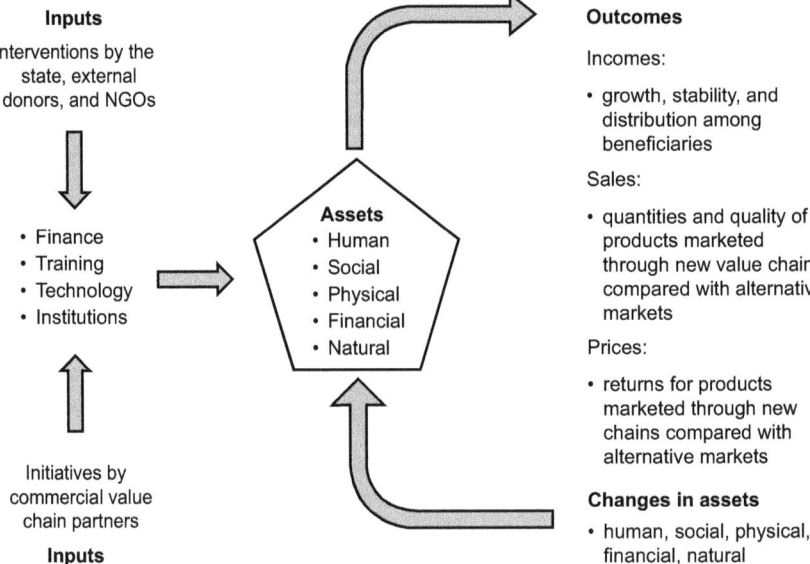

Figure 5.1 Theory of change approach: intervention inputs, assets, and outcomes
Source: author

change results in an increase of livelihood assets (natural, human, social, physical, and financial), and asset changes have been found to be useful elsewhere as indicators of livelihood change (Donovan and Poole, 2013, 2014).

The following key asset indicators were analysed for the four projects:

Human assets:

- improved technical skills for production and managerial skills;
- improved quality-control protocols;
- development of skills in new product development and exploitation of new markets;
- number of new product lines.

Social assets:

- access to markets and information;
- participation in collective activities – product sales through intermediary organizations;
- participation in collective activities – organizational and membership and governance activities;
- formalization of contractual linkages with value chain intermediaries.

Physical assets:

- technology, buildings, equipment, machinery, and housing improvements;
- investments in processing and marketing infrastructure – individual and/or collective.

Financial assets:

- access to credit;
- income benefit from product sales.

Natural assets:

- scale of production, increases in livestock numbers, or new land brought into production;
- uptake of new production technologies such as new varieties;
- investments in resource conservation and management.

Market-related indicators

The second set of indicators relate to market participation by the smallholder farmers:

Incomes:

- changes in producer/beneficiary incomes;
- changes in the distribution of income between men and women (likely to be derived from qualitative data).

Sales:

- changes in total product sales (quantity);
- proportion of product sales through new market outlets, compared with 'traditional' or alternative (status quo ante) outlets.

Prices:

- product prices received through new market outlets compared with 'traditional' outlets;
- prices and income stability.

Innovation:

- proportion of income from new economic activities, compared with traditional activities;
- new income-earning opportunities e.g. labour markets, associated services.

Methodology

Data requirements

The assessment entailed the following key steps:

- review of documents and literature;
- development of an analytical framework that included the design of instruments for the different categories of interviewees/respondents, namely project beneficiaries and stakeholders;
- development of a work plan for country field visits in Peru, Ethiopia, Tanzania, and Zambia;
- field visits to the four countries for data collection, with each country allocated five days;
- administration of individual questionnaires to project beneficiaries and in-depth interviews with project stakeholders;
- site visits and observation of project facilities and activities;

Triangulation was emphasized in the data-collection process. A detailed account of the approach and methodology can be found in Mudungwe et al. (2012).

The main data-collection instruments and methods employed in data collection comprised interviews with project beneficiaries based on structured questionnaires and in-depth interviews with project stakeholders.

The assessment questions were designed to capture both quantitative and qualitative data. The instruments captured respondents' personal data (age, sex, education, household status) and changes in livelihood assets (household assets, agricultural assets, social assets, financial assets – access to credit and income gains). The instruments also captured the extent to which specific constraints to market participation by smallholder farmers were addressed by project interventions.

Table 5.1 Respondent categories and numbers

Country	Project beneficiaries	Stakeholders	Total
Ethiopia	23	8	31
Peru	29	10	39
Tanzania	40	14	54
Zambia	40	16	56
Total	132	48	180

Convenience sampling and purposive sampling were employed for the assessment in selecting individuals in the target groups. A major determining factor was accessibility of the target groups within the time frame of the visits. Table 5.1 shows the number and categories of respondents in the four countries.

Field methods

Field visits were made to each project site during 2012. The participatory techniques involved qualitative data collection through close collaboration between the evaluation team and project funders, supervisory bodies, executing agencies, and a sample of project beneficiaries. Discussions were held with market-chain actors such as private-enterprise executives from trading and processing firms, cooperative leaders and members, individual farmers and project participants; and with key informants such as project managers, and public-sector stakeholders.

Data analysis

There was both qualitative and quantitative analysis of data collected for the assessment.

Content analysis was used in analysing qualitative data from interviews and focus groups, and consisted of identifying themes, issues, trends, and thematic interrelationships. The technique involves looking at documents, text, or speech to see what themes emerge, or what people talk about the most.

Descriptive statistics were used to describe relationships between variables and to suggest any causal relationships between variables being analysed. Frequency tables, where statistics such as simple percentages are used to compare before and after situations, were used.

Summary quantitative analysis using regression techniques was also undertaken. Findings have been incorporated into the conclusions below.

Limitations of the studies

The four studies presented in Chapters 6–9 are not necessarily representative of all the projects supported by the programme, and summarize only the issues and factors that are relevant to the theme of market participation.

Nevertheless, the research approach, concepts, methods, and the theory of change are amenable to further development and will have more widespread application.

A number of challenges and constraints were encountered in conducting the assessments. These included:

- insufficient time (days) to enable a wider and more comprehensive coverage;
- unavailability of some key documents for review (for example a project-appraisal report for Tanzania) before conducting the field research;
- difficulty in accessing some of the key stakeholders and beneficiaries;
- choice of respondents was restricted to those within a reasonable distance – in view of the time allotted those in distant places, especially remote beneficiaries, could not be accessed;
- language difficulties and difficulties some beneficiaries had in understanding concepts made it difficult to collect complete data on yield, quality, and income changes;
- rescheduling of appointments (for example the Zambia field visit) resulted in delays in completion of the data-collection component.

Summary of lessons learned

For those who want an overview, this synthesis of lessons contributes to advancing the design of project interventions, and guiding investment by commercial agribusiness firms that source supplies from smallholders, so that they more effectively target different categories of smallholders:

- The commodity projects analysed mostly targeted better-off smallholders, those with relatively better access to productive assets, and those farming under more suitable agroecological conditions. This bias towards the better-endowed smallholders almost certainly generated the best economic returns to the intervention, but it implies that the more marginal people in less-favoured contexts were likely to be excluded as beneficiaries.
- Targeting matters. Resolving the contextual ethical issues requires consideration of equity among different potential participants, and the ability of potential beneficiaries to engage in higher-risk activities. Is targeting the better-off a deliberate policy to get the best returns, or an easier way to achieve outcomes?
- Improvements in smallholder market participation were associated with project activities that focused on extension, training and demonstrations, and support for building up private agricultural assets. Market participation was also correlated with initial conditions related to household and farm characteristics such as wealth, land size, asset ownership, and prevailing agroecological environment.

- Access to credit was found to significantly influence farmers' access to market, highlighting the positive role of credit-support activities which constituted, in several instances, a core component of the projects.
- Given the existing bias in the selection of participating smallholders, project activities and interventions need to be specific to the targeted group. For the better-off smallholders, priority should be given to activities addressing standards, quality, and export markets which, from the field surveys, appear to be one of the main barriers to expanding market access. For poorer smallholders, priority should be given to activities that build up private productive assets and access to financial services, minimize risk, and preserve natural capital.
- This suggests that project design should consider realistically that desirable outcomes for different types of smallholder farmers – better-off versus worse-off – may not be the same; participants' aspirations may differ, with risk aversion being one significant variable affecting farmers' responses (Poole et al., 2013).
- Results also suggest that projects that specifically address aspects related to the natural, human, social, physical, and financial capitals that determine the livelihood opportunities of smallholders are likely to contribute significantly to strengthening market linkages.
- Projects that included on-farm risk management strategies were more successful because they took into account the wider agroecological, social, and economic environment.
- Similarly, interventions including elements that provided for fair, clear, and balanced counterpart-contribution arrangements among project stakeholders, and a deliberate investment in value chain linkages within the context of a deep understanding of the wider market environment, stood the greatest chance of seeing gains in market access sustained beyond the lifetime of the project.

References

Amrouk, El. M., Poole, N.D., Mudungwe, N. and Muzvondiwa, E. (2013). *The Impact of Commodity Development Projects on Smallholders' Market Access in Developing Countries: Case studies of FAO/CFC projects*. Rome, United Nations Food and Agriculture Organization. Retrieved 28 March 2017, from http://www.fao.org/docrep/017/aq290e/aq290e.pdf.

Donovan, J. and Poole, N.D. (2013). Asset building in response to value chain development: lessons from taro producers in Nicaragua. *International Journal of Agricultural Sustainability* **11**(1): 23–37 <http://dx.doi.org/10.1080/14735903.2012.673076>.

Donovan, J. and Poole, N.D. (2014). Changing asset endowments and smallholder participation in higher value markets: evidence from certified coffee producers in Nicaragua. *Food Policy* **44**: 1–13 <http://dx.doi.org/10.1016/j.foodpol.2013.09.010>.

Harper, M., Belt, J. and Roy, R., eds (2015). *Commercial and Inclusive Value Chains*. Rugby, UK, Practical Action Publishing.

Mudungwe, N., Muzvondiwa, E., Musarurwa, G. and Solange Sevilla, S. (2012). *Assessment of the Impact of Selected Commodity Development Projects on Smallholder Participation in Agricultural Markets/Value Chains: Case studies from Ethiopia, Peru, Tanzania & Zambia*. Rome, Practical Action Consulting for Food and Agriculture Organization of the United Nations.

Poole, N.D., Chitundu, M. and Msoni, R. (2013). Commercialisation: a meta-approach for agricultural development among smallholder farmers in Africa? *Food Policy* **41**(August): 155–165 <https://doi.org/10.1016/j.foodpol.2013.05.010>.

CHAPTER 6
A diversification programme for vegetable exports in Ethiopia

This study examines a project designed to boost smallholder productive capacity and participation in the export chain for green beans from Ethiopia to Europe. Some impacts were positive. Overall, the project was unsustainable for reasons of poor project design and implementation, specifically the failure to create a viable value chain.

Keywords: export orientation, asset building, marketing, value chain, project design and implementation, Ethiopia

Project context

Programme design and implementation

Europe has long been seen as a potential market for exports of fresh produce from sub-Saharan Africa, although the challenges facing suppliers, particularly in terms of value chain organization and quality control, have long been evident (Jaffee, 2003). Evidence from a range of regions and countries shows the potential and constraints of smallholder participation (Blandon et al., 2009; Barrett et al., 2012). In Ethiopia, the participation of smallholder farmers in marketing cooperatives has been found to be likely to reduce rural poverty and increase agricultural commercialization (Francesconi and Heerink, 2011), although welfare impacts on the poorest of agricultural exports are not always as expected (Cramer et al., 2016).

The programme analysed here was designed to strengthen the export capacity of vegetable farmers in Ethiopia and Sudan through the removal of critical supply-side constraints and weaknesses in relation to technical, infrastructural, and market factors. The project was implemented over three years from May 2007 to April 2010. There was a further no-cost extension phase from May 2010 to July 2011.

Here, only project implementation in Ethiopia is considered, and we focus on four main components. The major outputs/activities of the project intervention were:

Component 1: Development of an integrated export programme for green beans

- identification of production area and producer profiles;
- market-oriented production-planning meetings with stakeholders along the value chain;

- facilitation of a three-year marketing agreement for export production between farmers' union and exporter.

Component 2: Development of export-oriented green-beans production system

- training of farmers and extension workers on land management, crop production, harvesting, and post-harvest handling technologies;
- production of manuals and posters for farmers and extension workers;
- creation of a credit-revolving fund with farmers' unions to assist in the purchase of seeds;
- establishment of a safe and prescription-based plant protection system;
- establishment of a traceability system from pack-house to the exporter system, employing records plot, farmer, picker, water-user group, grader at pack-house, dispatch date;
- training and creating awareness on Global Gap, an international quality assurance system.

Component 3: Post-harvest handling and quality assurance from farm gate to export point

- training on HACCP and on quality-control systems for pack-houses;
- two pack-houses constructed and operated by the project for two seasons.

Component 4: Development of marketing and trade-facilitation services

- study of the existing market-information systems and options for development undertaken in collaboration with Wageningen University;
- establishment of chain linkages between producers and the exporter.

Questionnaire sample

A total of 23 farmers were interviewed. All were male, and all were heads of family with the exception of one who was the son of the head of family. All households considered themselves to be middle-income. Relevant personal and project participation data are presented in Table 6.1. More than half of the farmers included in the sample were over the age of 30, and four over 50 years. Almost half had high-school or diploma-level education. Half of the sample had been with the project for three years, and half for only one year.

Table 6.1 Personal data and project participation

Age (years)		Level of education		Date of organization membership		Years of participation in project	
25–30	10	Primary	12	Pre-2000	7	1	10
30–50	9	High school	8	2000–07	10	2	0
>50	4	Diploma	2	Post-2008	3	3	10

Findings

Household asset building

Respondents were asked to indicate what significant household assets they had before and after the project. New or increased income was likely to have been spent on recurrent expenditures such as food and clothing. Responses were classified as a 'minor upgrade' if the item was a consumption good or household item (radio, bicycle, TV, household furniture). Each item was probably worth less than 1,000 birr (US$43; see note at end of chapter). A total of five households were identified as having made no minor investments or asset upgrades during the life of the project. A total of 18 households, or 78 per cent of the sample, acknowledged increases in minor household assets.

A response was classified as a 'major investment' if the item was a productive asset (motorbike, at least one ox or cow, a cart, or investment in improvements to housing), the cash value of which was likely to exceed 1,000 birr for each major investment. Six households reported no major investments in household assets, 13 reported one major investment, and four reported two or more major investments.

Access to land

Land tenure was a complicated issue, involving own land, rented land, and share-cropping arrangements. These details were not investigated. The mean land area devoted to agricultural production initially was 3.3 hectares, rising to 4.1 hectares at the end of the project. There was considerable variation in scale of agricultural production, but many households were able to increase the scale of production during the course of the project. None of the respondents had grown green beans before the project, and the area under green beans was a small proportion of the total under agricultural production.

Changes in agricultural assets

Respondents were asked to indicate what significant agricultural assets they had before and after the project. Responses were classed as minor, moderate, or major upgrades (Table 6.2).

Table 6.2 Agricultural asset changes

Changes in assets	Number of households
Negative	2
None	7
Minor: up to 5 sheep/goats	2
Moderate: up to 10 sheep/goats/1 bovid (cow/ox)	4
Major: 2 or more bovids (cattle/oxen)	8

It is notable that two households suffered decreases in agricultural assets. It should not be a surprise that some households report negative asset flows. This was in part due to substitution of asset classes, but also due to losses incurred by engaging in what is essentially a risky economic venture. Two households balanced a reduction in livestock with an increase in major and minor household assets, or an increase in household assets and green-bean production. In both cases, this involved a shift from livestock into horticultural production, demonstrating considerable entrepreneurial capacity to manage a wider portfolio of activities. The labour requirement for horticultural production is a plausible reason why smallholders need to rebalance their asset portfolio. The asset reduction therefore was only nominal – until, probably, the loss of the export market rendered the investments in green-bean production void (see below).

Project services

All respondents acknowledged receiving services of agricultural extension, training, and access to inputs by means of the revolving loan fund. When asked about the adequacy of these particular services, results were positive: almost all responses were that services were either adequate or very adequate. A paired t-test comparing the responses 'inadequate' with 'adequate/very adequate' showed a difference that was very highly significant.

Responses related to the specific content of production services received were also generally positive but there were more responses affirming inadequacy. Nevertheless, a paired t-test comparing the responses 'inadequate' with 'adequate/very adequate' showed a difference that was highly significant ($p = 0.0038$). These results were supported by qualitative data from a smallholder respondent: 'We got knowledge on proper use of inputs, and my income has increased.'

One key informant commented: 'The project success was on the production side – given the right skills, small farmers can produce ...' (Ministry of Agriculture respondent).

The project design addressed satisfactorily the issues of productive capacity – inputs and technical issues – and the productive skills of the growers. From being non-growers to successful growers and exporters of green beans within three years (in some cases within one year) was a considerable achievement: 'Training received from the project helped us to use resources wisely and in a timely fashion' (smallholder farmer). 'The intervention was appropriate – it did build capacity for farmers' (farm manager, commercial producer/exporter).

In contrast, very few respondents considered any of the marketing services to be very adequate, and a considerable number considered the services to be inadequate, notably in respect of market information, new markets, and market linkages (Table 6.3). A paired t-test comparing the responses 'inadequate' with 'adequate/very adequate' showed a difference that was not significant at the 90 per cent level ($p = 0.1319$):

Table 6.3 Adequacy of marketing

Service	Inadequate	Adequate	Very adequate
Quality of produce	2	14	2
Market information	7	10	0
New markets	9	7	0
Linkages	8	9	0
Supply consistency	3	13	1
Transport costs	3	10	3
Contract enforcement	2	10	4
Prices	5	13	3
Storage facilities	0	0	0
Suitable transport	3	13	2

Note: * Difference of means between 'inadequate' and 'adequate/very adequate' not significant at the 90 per cent level ($p = 0.1319$)

Financial assets

Questions about savings and actual income gains were too complex to gain useful information because of translation and conceptual problems. Nevertheless, the positive changes in income and assets noted were attributed to the impact of the project. Qualitative data were more helpful than the quantitative data. A smallholder respondent said: 'The project linked us to market and we got higher prices and therefore increased our incomes'.

In addition, in respect of new income-earning opportunities, qualitative responses stated that the building of the pack-houses created additional temporary employment opportunities in construction and also seasonal work (which proved to be temporary because of the demise of the market). Also, there was a growth of the market for transport through use of donkey carts.

Downstream activities

However, there were major weaknesses in the project. Most importantly, the export of green beans was stopped by the commercial producer/exporter after the three-year project finished, and therefore the project failed completely to integrate smallholders into the export market in a sustainable manner. Farmers were left with no export market and there never had been a domestic market for green beans.

There was evident dissatisfaction with downstream activities: 'The supply chain should be addressed in an integrated manner (production, harvesting, grading, cold chain management, packing) ...' (key informant from project-executing agency).

Thus, the project badly failed to address marketing issues. This dissatisfaction was highlighted in interviews with smallholders, many of whom commented to the effect that:

> The project should have assisted us with creating a sustainable marketing network, thereby increasing our incomes.

> The project should have linked us also with local buyers.

> The project should have created a sustainable market-information system to aid farmers in production decisions.

Sustainability

As the commercial producer/exporter stopped trading green beans at the end of the project, the sustainable integration of the smallholders was unsuccessful. As one key informant commented: 'Farmers in Ethiopia are no longer producing green beans because the exporter they were linked to is no longer commercializing green beans' (project-executing agency).

The negative comments are in response to the closure of the market at the end of the project, even if the marketing during the project had been considered successful:

> Farmers were trained in market-oriented production planning where there was no market sustainability – an area of concern is the lack of attention given to sustaining the project initiative and the genuine innovations (project-executing agency).

> Market sustainability was not addressed by the project (key informant – Wored, Ministry of Agriculture Head, Ziway, Oromia Region).

Moreover, for some of the smallholder growers, the short-term benefits of exporting will have been outweighed by the losses incurred through the diversification investments in horticulture which became sunk costs. In short, some growers will have been impoverished by participating in the project.

Costs of quality

Another issue of direct relevance that surfaced in discussions was the high-cost and demanding quality-assurance criteria for the European market. Much is known of the challenges this poses to exporters in developing countries. For smallholder cooperatives to gain and maintain access to European markets without the intermediary services of a major commercial exporter requires a different level of investment in capacity building and financial operation. The conclusion drawn from the focus group discussions was that: 'There is a need for continuous training on production, packing and market information'. And, from a key informant: 'The high cost of the certification is seen as challenge for small farmers' realization of quality-assurance certification' (project-executing agency).

Targeting

Another significant issue was the targeting of smallholders. All those interviewed claimed to be middle-income farmers. The initial level of household assets supported this view. It is questionable whether the poorest farmers were – or indeed should have been – included. This approach towards the better-off was endorsed by one key informant: 'Promoting and targeting middle-income people is a good strategy – it provides a field laboratory where others can learn – also there are fast livelihoods improvements' (project-executing agency).

So the impact on wider poverty reduction and smallholder inclusion is indeterminate. While the analysis has focused on smallholder farmers, there was some dissatisfaction from the key informants within the cooperatives that insufficient capacity building was directed at the primary cooperatives and the second-tier cooperative which supplied the exporter: 'The project could have built the capacity of the cooperative union by setting up market-intelligence units' (Ministry of Agriculture, Oromia Region).

Project planning and implementation

The lack of sustainability beyond the three-year horizon was a function of poor project design and implementation. Various issues arising from the project background are worth highlighting, and not all the elements of the project were completed satisfactorily. In particular, the failure to address four issues increased the level of risk of the project:

1. Because export-quality horticultural produce is usually a small proportion of overall production, 'it was advised that importance should be given to the promotion of regional and possibly also domestic markets for produce which do not meet export standards' (para 3). The regional dimension to the market possibilities in the project documentation is evident, but in Ethiopia the focus was entirely on the export market, because green beans were not (much) consumed locally. As key informants commented:

   Beans were a new crop to the farmers and had no local market (project executing agency).

   Regional markets could have been explored, for example, Djibouti, Saudi Arabia ... (local team leader, Irrigation Agriculture Office, Oromia Region).

   There is a short export market in Europe so there is a need to diversify to regional and local markets (farm manager, commercial producer/exporter).

2. The project envisaged trying three crops, and possibilities need not necessarily have been confined to these. Green beans and okra were selected for the initial phase of the project, with possible additional crops

(e.g. capsicum peppers) to be introduced in the third year. In Ethiopia, only green beans were grown, increasing the risks significantly.
3. The project documentation highlighted the importance of efficient value chain linkages, making recommendations about the contractual arrangements between producers and exporters. Three commercial partners were initially identified. However, there was a close family link between the project-executing agency and the only commercial grower/exporter involved in the project. Such a single-channel market structure was unlikely to be efficient and competitive, and, evidently, was unsustainable. Whereas the project-executing agency was noted as encouraging overseas investors including Dutch, German, Israeli, UK, and US horticultural companies to consider investment in the sector, it does not appear that foreign direct investment was considered to be a source of capital and expansion for this project, nor, most importantly, for creating more value chain linkages to diverse export markets. 'The weakness of the project was dependence on one market – there was a need for market diversification – there was a need to link with the private sector in market development' (local team leader, Irrigation Agriculture Office, Oromia Region).
4. Finally, it is evident that a counterpart contribution was envisaged, and indeed was considered to be important to secure commercial participation. The counterpart contribution should have been quantified and guaranteed. It is not clear that this was done, which is surprising considering that the 'counterpart' was a major export business.

It is evident that risk was increased by the exclusive focus on the export market, on one single crop, and on one single export organization in the value chain. Moreover, risk was magnified through an inadequate counterpart contribution which enabled the exporter to cease trading at the end of the project without the incurring the penalty of sunk costs, leaving producers with no market outlet. We can speculate likewise that the lack of competitiveness in the new value chain can be attributed to the close relationship between the project-executing agency and the commercial producer/exporter.

Conclusions

Project beneficiaries acknowledged that there had been income and livelihood improvements which they attributed to the project, that they were able to increase the scale of agricultural production overall, and in most cases to build assets. Capacity building among farmers was a success: significant skills in production were gained. However, the risks of asset switching into green beans were high, and at least some of the assets have become redundant; smallholders who invested in assets specific to green-bean production and exporting will have incurred significant sunk costs. Therefore, some growers may actually have been impoverished by participating in the project.

The demand for agricultural labour increased to meet the need for careful picking of green beans, and there was a new demand for labour in the construction and operation of the pack-houses. The acquisition of carts by some farmers opened up new economic opportunities as transporters.

Institution building was unsuccessful: the failure to build viable value chain linkages was a major negative outcome of the project that imperilled sustainability: there were no enduring chain linkages and business networks established between producers, cooperatives, and sustainable export partners. The loss of the export channel meant that few of the direct benefits have been sustained. The unresolved question of counterpart funding may be significant in determining how and why the business closed on completion of the project.

Summarizing: three major strategic mistakes were the exclusive focus on the export market, on one single crop, and on one single export organization.

Note

The 1,000 birr heuristic used to distinguish between minor and major investments is a matter of judgement, and it is arguable that a cart and a water pump, for example, should not be classified as minor investments but as significant productive assets (U$1 = 23 birr on 24 May 2017).

References

Barrett, C.B., Bachke, M.E., Bellemare, M.F., Michelson, H.C., Narayanan, S. and Walker, T.F. (2012). Smallholder participation in contract farming: comparative evidence from five countries. *World Development* **40**(4): 715–730 <http://dx.doi.org/10.1016/j.worlddev.2011.09.006>.

Blandon, J., Henson, S. and Cranfield, J. (2009). Small-scale farmer participation in new agri-food supply chains: case of the supermarket supply chain for fruit and vegetables in Honduras. *Journal of International Development* **21**(7): 971–984 <http://dx.doi.org/10.1002/jid.1490>.

Cramer, C., Johnston, D., Mueller, B., Oya, C. and Sender, J. (2016). Fairtrade and labour markets in Ethiopia and Uganda. *Journal of Development Studies* **53**(6): 841–856 <http://dx.doi.org/10.1080/00220388.2016.1208175>.

Francesconi, G.N. and Heerink, N. (2011). Ethiopian agricultural cooperatives in an era of global commodity exchange: does organisational form matter? *Journal of African Economies* **20**(1): 153–177 <https://doi.org/10.1093/jae/ejq036>.

Jaffee, S. (2003). *From Challenge to Opportunity: Transforming Kenya's fresh vegetable trade in the context of emerging food safety and other standards in Europe*. Washington, DC, World Bank.

CHAPTER 7
Production and commercialization of oilseeds in Peru

This study examines a project designed to develop the oilseed sector in Peru through encouraging smallholder production of jatropha (Jatropha curcas). Some impacts were positive, notably human- and social-capital formation. The case provides evidence that creating a viable value chain takes time, particularly for agricultural crops that have a significant industrial development component, and which are intended to address multiple objectives such as poverty reduction and enhanced environmental management. The complexity of designing multisectoral interventions in production, processing, and transport was not recognized. For assessing economic improvements, a longer trial period is necessary.

Keywords: jatropha, technical services, sustainability, value chain development, processing, research, Peru

Project context

Programme design and implementation

The development of the oilseed sector in general, and of jatropha in particular, offers opportunities for smallholder participation in the bioenergy industry (Brittaine and Lutaladio, 2010). Nevertheless, linking smallholders into such markets requires investment into more than production and marketing: significant research requirements and interaction with public energy policies presuppose coordination with the state and private enterprises, plus patient capital investment. Mixed experiences mean that it is no longer considered to be a 'miracle' crop (Degail and Chantry, 2013).

The project in Peru and Honduras started in 2007 and was due to finish in 2013. Working with small-scale agricultural cooperatives, it was designed to reduce poverty and to improve the quality of the lives of the farmers by improving the supply of and demand for rapeseed and *Jatropha curcas*. The project's aim was to provide biodiesel as an alternative to conventional fuels in Peru. The seeds would also be processed for the production of vegetable oils as a substitute for diesel fuel in public transportation. Relevant government institutions were part of several inception meetings to promote the planting and marketing of jatropha.

The project was complex and, to some extent, experimental, with five components and outputs. Essentially it was a private–public partnership between donors, smallholder farmer groups, and small-scale companies in Peru and

Honduras to develop renewable energy alternatives, increase employment opportunities, and to contribute to poverty reduction and livelihood improvement.
Here we report only on the case in Peru.

- *Component 1: Agricultural development.* Activities required for participating farmers to successfully cultivate jatropha and rape for seed production.
- *Component 2: Production.* Producing the natural vegetable oil.
- *Component 3: Public transportation.* Uptake by private-sector commuter bus companies of vegetable oil fuels as a substitute for diesel fuel.
- *Component 4: Training.* Activities for the main stakeholders involved in the project.
- *Component 5: Promotion and dissemination.* Information on using natural vegetable oil technology including the production of rape and jatropha, the extraction of vegetable oil, and the conversion of diesel motors.

Demographic details

The project was implemented in four regions of Peru, and this study was undertaken in districts of the Lambayeque region of the maritime north. Twenty-nine (29) farmers were interviewed. The project was dominated by men, with only three women as participants, reflecting the prevailing socio-economic organization which has low participation of women in decision-making positions for agricultural production and marketing.

The majority of farmers interviewed were aged 30 to 50 years and most of them had completed secondary-school education. All but two were heads of household. Details are shown in Table 7.1. The high level of education of some of the participants (seven with tertiary education) was notable. Most farmers were well-established in the area, with decades of residence and up to five years of cooperative membership.

Findings

Not all quantitative data were complete, and data such as distance to market were not informative, giving only a partial insight into the implementation

Table 7.1 Personal data and project participation

Age category (years)		Level of education		Years living in the area		Relationship to household head (HoH)		Date of cooperative membership	
25–29	7	Incomplete	1	9–20	2	Child	2	2007	13
30–50	17	Primary	4	21–30	5	HoH	27	2008	1
>50	5	Secondary	17	31–40	9			2009	2
		University	7	41–50	8			2010	4
				>50	5			2011	7
								2012	2

and impact of the project. Here we will concentrate on the qualitative findings in particular, which give important insights into project design and implementation.

Scale of agricultural production

Jatropha was a recent introduction to the project area. prior to the project there was small-scale production but no farmer had yet been able to harvest the product, so there were no productivity data available. The introduction of the project saw farmers' land under jatropha increasing from a mean of 1.8 hectares to 3 hectares, with a maximum of 18 hectares. This growth was attributed to new, larger-scale entrants into jatropha farming as well as increases in area by incumbent farmers.

The introduction of the project saw farmers harvesting an average of 1.9 metric tonnes per hectare, with a maximum of 4 tonnes per hectare. Some of the new farmers were yet to harvest at the time the survey was undertaken.

Project services

All respondents acknowledged receiving services of training, agricultural extension, and access to inputs. When asked about the adequacy of these particular services, results were positive: almost all responses rated them as either adequate or very adequate (Table 7.2).

Despite these positive responses, many farmers commented about constraints to productive capacity and the limits to information.

Productive capacity

Farmers noted that irrigation systems were crucial for increasing productive capacity. However, in some of the project areas they faced acute shortages of irrigation water. There were failures in the first plantations of jatropha that were not irrigated and were inadequately fertilized owing to a lack of technical information. Some had water reserves in wells approximately 80 metres deep but problems were compounded because they did not have adequate pumping facilities to extract the water from such deep wells.

Jatropha achieves optimum production in its sixth year, and this is the period when farmers start to recover their investment. Most of the crop in the project area was in its fourth and fifth year, and yields were still low, the maximum benefits having not yet been realized.

Table 7.2 Responses concerning adequacy of services provision

	Training	Extension	Inputs
Inadequate	2	3	2
Adequate	24	15	22
Very adequate	1	1	1
No response	2	10	4

Regarding financial returns, farmers complained that the sole company purchasing their crop was paying low prices. Lack of investment was also a constraint. The following comments highlight the farmers' own perceptions of the problems arising from physical infrastructure failings and technical/technological support:

> We need financial support for the installation and use of irrigation technology.

> We need to improve the irrigation system, develop wells, and improve access to energy for processing.

> We need pipes to transport water to the farm, we need pumps to extract water from wells, we also want to learn how to use and manage pesticides.

> We need fertilizers, training, better tools, pumps to move water, irrigation technology for increased production.

> The seed production has dropped due to lack of water, we got an electric water pump but we need help to install it.

> We need assistance with purchasing equipment as well as marketing our product.

> We get little seed from the project; we need more seed to increase production to be profitable.

Knowledge management and technology

Misinformation and lack of training were mentioned by participants as constraints. From the beginning of the project farmers used one type of seed that is common in the area. This single variety is usually attacked by fungal diseases that reduce yields. Farmers indicated that they had been made to believe that jatropha was not prone to pests and drought. Experience pointed to the contrary; many farmers also experienced pests which attacked the crop.

Unrealistic expectations undermined the confidence of the farmers: 'When the project started we had false expectations that jatropha was not affected by pests. When the pests started to attack the crop, a lot of farmers were disappointed and abandoned the crop.'

Farmers were also disillusioned by the poor returns compared with the labour demands of the crop.

> The price at which we sell the jatropha does not represent all the work required and the money spent to make it grow.

> The price of jatropha needs to be increased to cover costs of everything that is done in the field (inputs and labour costs).

At the time of the research, it was evident that a small percentage of farmers were likely to pull out of the project if support were not forthcoming for necessary infrastructure, irrigation, fertilizers, harvesting, and market access.

Sustainability

Since it takes five to six years for jatropha to be ready to harvest, it is necessary for farmers to have other seasonal crops to assist with income generation while waiting for the crop to mature. Doubts about the appropriateness of jatropha for prevailing livelihood systems were suggested by a key informant: 'The cultivation of jatropha among the farmers is not very desirable because one must wait five years to see a proper return to make profits' (key informant, agro-processing).

Farmers made wide-ranging comments about the introduction of jatropha into their productive systems. For example: 'We need to learn to improve agricultural production of other crops such as squash and bell pepper. Farmers also need to consider jatropha since it generates income that helps the family budget.'

Removing project support was said to be likely to create risks that farmers would abandon their crops, since they could not absorb all the production costs alone.

Public support

Farmers also criticized the lack of institutional development in terms of overarching government strategy and post-harvest development activities for the sector:

> There is need for greater involvement of the Ministry of Agriculture to continue to support farmers through extension services.

> There is need to further investigate how to increase extraction of oil to more than 30 per cent from the oil cake.

> We need to disseminate information on the use of biopesticide for crops because it is an organic product.

> There is need to investigate how to detoxify the cake of jatropha for use as stock-feed for cattle.

> More research is needed on the cultivation of jatropha; you also need more research on the cake of jatropha and the use of latex.

A key informant from the regional agriculture directorate added: 'The project should be strengthened to maximize the capacity of farmers through training, incentives, financing'.

Interviewees reported inadequacies in the package of technologies for handling pests and diseases, and also commented that skills in negotiating and marketing still needed to be improved. The long-term nature of investment in the production of jatropha was highlighted by another key informant from the public-sector national agricultural research institute: 'It is too early for the project to end, its life span must go up to at least ten years, so that it sees farmers through one harvest.'

Impacts

Strengths

The project had a series of positive impacts in terms of human-capital creation among farmers and in the formation of social-capital value chain linkages. Farmers learned new techniques of management of jatropha including pruning, use of organic manure, biopesticide use, and intercropping short-term cash crops with jatropha to sustain incomes during the investment stage of the crop. Jatropha was also adopted for the development of live fences.

There was important evidence of value chain development in terms of new products, market destinations, and value addition. The biopesticide niche market for jatropha was starting to look very attractive to entrepreneurs and farmers enabling them to find a wider market for the seeds. Another important development was the farmers' own manufacture of organic fertilizer called BIOL made from elements found easily on the farm. Nevertheless, the commercial processor was also dedicated to marketing the oil and marketing the by-products for other uses such as the growing market for biopesticide using the plant's properties as a fungicide and insect repellent. It has also encouraged farmers to start developing their own organic fertilizers.

Farmers' entry into beekeeping was reported to have resulted in significantly increased crop yields through better pollination. The diversification also introduced a new source of income and livelihood for farmers who could both consume and sell the honey.

Farmer associations were strengthened and new organizations were formed, for example the Association of Agricultural Producers Motupana Piñon (ASAMPRO) and the Agricultural Producers Association of Monteria in Lambayeque. These were formed by 32 farmers who were active participants in the project. Social interaction was said to have improved because farmers attended agricultural-association meetings.

Interviews also suggested a number of benefits through the reinvestment of income in household assets, minor improvements in housing infrastructure, and the acquisition of new physical agricultural assets. In Lambayeque, respondents said they had not acquired any assets over the duration of the project but improvements to the homes of farmers were known to have been made and some farmer participants also bought new properties. There were positive impacts on the wider economy as the demand for agricultural labour to harvest jatropha seed also grew.

Weaknesses

For farmers, jatropha is one element among other livelihood activities – such as food production and other income generation – and is undertaken within an institutional environment which needs a higher level of broad-based sectoral support: competitive markets, accompanying technological development in production and processing, and appropriate information, research, and training.

A series of technical and project-design weaknesses was also evident. Farmers' comments highlighted the importance of considering the development of jatropha as just one part of an agricultural system, and production as one element of an integrated value chain with considerable downstream functions.

Another significant issue was the targeting of smallholders. Most farmers participating in the project in Lambayeque were from the middle-income stratum, some had parallel occupations, others were also traders and most were farming food crops for income generation.

Value chain linkages emerged as a result of the project, but were not entirely successful. An agreement with a public-transport company was made, leading to the purchase of engine conversion kits for pure vegetable oil fuel. Unfortunately, technical problems developed with the conversions and there was limited demand for oilseed for vehicle fuel or for power generators.

In terms of market structure, monopoly and monopsony power was evident: at the time of research there was only one commercial company responsible for buying the seeds and undertaking processing for oil extraction and marketing of the biofuel. The processor was also the only source of seeds for planting.

Farmers were aware of the uncompetitive nature of sales and had demanded that the sale price of the seed be increased (by three times). They expected more of the cooperative and argued that there should be regulation setting the price for farmers to obtain at least 15 per cent profit after the sale of jatropha. The processor's market power was considered to affect the terms of the exclusive purchasing agreements: farmers were disadvantaged by the fixed seed price, and were aware of the risk that was transferred to the farmer rather than shared by the processor.

Conclusions

These findings were endorsed in a final project report (GIZ 2013). The report acknowledged that it was a risky venture 'characterized by high uncertainties and lack of knowledge about important issues, especially the development of the energy markets and the suitability of the jatropha plant as an economic viable alternative for small farmers' (GIZ, 2013: 3). The desire to provide participation opportunities to smallholder farmers in renewable-energy markets was admirable but the experimental nature of the project made the design and implementation imperfect.

The political and environmental objectives of the project have not been evaluated here. For all the complexities of the project, it is unsurprising that the sustainability of the value chain model was not in any way assured at the time of the analyses. The report acknowledges that the many local constraints to smallholder participation and small-enterprise development were too serious to be overcome by the project intervention; similarly the human-capital weaknesses, the novel mono-cropping approach, high costs of support to a small and fragmented farmer base, unequal risk-bearing, a series of technical production issues already covered, and the agro-industrial and market weaknesses. In a quote which summarizes the human capacity challenges: 'It is not possible to change from a subsistence economy mentality to a management mentality, if this process is not accompanied by ongoing advice and training of new skills' (GIZ, 2013: 16).

For our purposes, there are a number of specific lessons learned from the project design and implementation which could help future interventions:

- development of appropriate production technologies and infrastructure, some of which required further research before project implementation;
- delivery of comprehensive information, and appropriate knowledge extension to build the capacities and skills of project participants;
- analysis of the equity implications of the economic conditions of prevailing market structures;
- formulation of appropriate contractual or regulatory institutions to redress undue market power and unequal-bearing;
- a lengthened project cycle framework which acknowledges the extended time-frame for: (1) the development of human skills; (2) sectoral growth towards maturity of supply and demand; and (3) growth of the industrial demand and bio-transport sector.

Similar challenges have been observed in other, related projects (Poole, 2010; Poole et al., 2010), which add weight to the imperative to rethink project-design processes long before project implementation.

Overall, this was a complex programme, with multisectoral interventions in production, processing, transport, capacity building, and market development. While it was helpful to recognize the integrated nature of the necessary components for the project to succeed, the challenges exceeded the capacity of the implementing organization.

The relatively short time frame of the programme was insufficient to enable sustainable changes, which is a common failing in social-development projects in Latin America and elsewhere (Donovan et al., 2017).

References

Brittaine, R. and Lutaladio, N. (2010). *Jatropha: A Smallholder Bioenergy Crop. The Potential for Pro-Poor Development. Integrated Crop Management Vol. 8*, International Fund for Agricultural Development/Food and Agriculture Organization of the United Nations.

Degail, A.-C. and Chantry, J. (2013). Developing jatropha projects with smallholder farmers: conditions for a sustainable win-win situation for farmers and the project developer. *Field Actions Science Reports [Online]* (Special Issue 7). Retrieved 23 May 2017, from https://factsreports.revues.org/2182.

Donovan, J., Blare, T. and Poole, N. (2017). Stuck in a rut: emerging cocoa cooperatives in Peru and the factors that influence their performance. *International Journal of Agricultural Sustainability* **15**(2): 169–184 <http://dx.doi.org/10.1080/14735903.2017.1286831>.

GIZ (2013). *Production of Oleaginous Plants and Commercializatiion of Natural Vegetable Oils as Substitutes for Diesel Fuel for Public Transportation in Peru and Honduras (CFC/FIGOOF/26)*. Retrieved 18 November 2016, from http://common-fund.org/fileadmin/user_upload/Projects/FIGOOF/FIGOOF_26/1.Technical_Report_Project_FIGOOF_26.pdf.

Poole, N.D. (2010). *Zambia Cassava Sector Policy: Recommendations in support of strategy implementation*. EU-AAACP Paper Series. No. 16. Rome, Food and Agriculture Organization of the United Nations. Retrieved 21 February 2017, from http://www.fao.org/fileadmin/templates/est/AAACP/pacific/07_FAO_AAACP_Paper_Series16_Recommendations_Zambia_Cassava_Strat.pdf.

Poole, N.D., Chitundu, M., Msoni, R. and Tembo, I. (2010). *Constraints to Smallholder Participation in Cassava Value Chain Development in Zambia*. EU-AAACP Paper Series. No. 15. Rome, Food and Agriculture Organization of the United Nations. Retrieved 21 February 2017, from http://www.fao.org/fileadmin/templates/est/AAACP/eastafrica/FAO_AAACP_Paper_Series_No_15_Constraints_to_smallholder_participa%C3%83__1_.pdf.

CHAPTER 8
Sisal product and market development in Tanzania

This study examines a project that built on a legacy of colonial-era sisal-sector development in Tanzania. It was a complex project integrating various production, marketing, and product-transformation services. We focus here on the agricultural-development components. The design and delivery of project technical services and capacity building were criticized by participants, even though some participants acknowledged several livelihood improvements, and increases in the scale of production were noted. The challenges of building a complex value chain with significant processing demands and market development in an uncertain external environment were not addressed.

Keywords: asset building, processing, value chain, project design and implementation, external environment, Tanzania

Project context

Programme design and implementation

Sisal (*Agave sisalana*) is a robust plant which can grow under adverse environmental conditions and produces a fibre for transformation into products such as rope, twine, carpets, yarn, and newer products such as paper. The case for there being market potential for a sustainable, renewable natural resource is strong (Shamte, undated). The production of sisal in Tanzania dates back to the era of German East Africa in the 19th century. Historically a plantation crop, there have been efforts to increase participation by smallholders since the early 1960s and the potential contribution of sisal to smallholder farming and wider economic development has been recognized (FAO Committee on Commodity Problems, 2013).

The project was designed to develop new sisal and henequen products and build the market for these products. The project had a multi-pronged and integrated value chain development approach, embracing both agricultural improvement and industrial product-development, both of which were supported by appropriate research efforts.

Project activities were implemented in Tanzania and Kenya during the period January 1997 to December 2005. Both countries have large smallholder farmer sectors and long-standing aspirations to encourage farmers to engage in commercial markets and contribute to wider sectoral development in order to escape the 'trap' of the subsistence economy. The analysis here is based only on the research undertaken in Tanzania.

In the 1960s, Tanzania was the largest producer of sisal in the world, but the industry suffered declines in output in subsequent decades. The potential for this endemic crop remains, and the industry weaknesses of inadequate research and development, and poor marketing arrangements, have been recognized for some time (Kimaro et al., 1994).

The project had five main components, namely:

- *Component 1: Agricultural development.* Development of new sisal varieties and improvement of cultivation practices.
- *Component 2: Product processing.* Utilization of fibre-extraction waste for animal-feed production and for biogas and electricity generation.
- *Component 3: Knowledge and capacity development.* Research and development on sisal processing for pulpable fibre.
- *Component 4: Market development.* Market studies and trials to establish the demand for sisal pulp and to identify potential buyers of the products.
- *Component 5: Extension.* Dissemination of project results.

This analysis has focused on the first component of improving agricultural production by smallholder farmers of fibre for both traditional and new uses. It was expected that, if produced and processed at competitive prices, sisal would have new market opportunities, in particular as a reinforcement fibre in the large and growing paper-pulp market. The activities concentrated on varietal selection, farming systems for sisal, evaluation of best agricultural practices, and plant multiplication by meristematic tissue culture in order to reduce production costs, improve productivity, and increase returns to labour and capital. An intermediate output was the attempt to identify an appropriate smallholder sisal outgrower model scheme for large-scale sisal estates.

Findings

Questionnaire sample

A total of 40 farmers from Korongwe District in Tanga region were interviewed, 20 each from two commercial estates. Respondent categories and numbers are shown in Table 8.1.

All the farmers interviewed were heads of family except one female who was a spouse. Thirty-four households considered themselves to be middle-income.

Table 8.1 Respondent categories and numbers

District	Estate	Project beneficiaries			Stakeholders
		Male	Female	Total	
Korongwe					
	Hale	15	5	20	
	Mwelya	16	4	20	14
Total		31	9	40	14

Table 8.2 Personal data and project participation

Age category		Level of education		Date of organization membership		Years living in the area	
25–30	1	Primary	25	Pre-2000	2	<10	3
30–50	23	High school	7	2000–07	35	10–30	8
>50	16	Diploma	3	Post-2008	1	30–50	14
		University	2			>50	12

Three households considered themselves poor and the other three households very poor. Other beneficiary characteristics are presented in Table 8.2, illustrating the relatively advanced age of participants, relatively low levels of education, and long-standing presence in the region. Most respondents lived within five miles of a road and the produce collection point and a service centre. The most distant lived 15–18 kilometres away.

Household assets

Respondents were asked to indicate significant assets they had before and after the project. In assessing the magnitude of change in household assets, the responses were classified as follows:

- no change – no household asset purchases;
- minor change – minor purchase of consumption goods (TV, radio, bicycle);
- major change/upgrades – investments in one or more major assets (house, car, motor cycle).

A total of 14, or 40 per cent, of the respondents recorded major upgrades in household assets. Nine households (26 per cent) had minor increases in household assets while 12 households (34 per cent) had no asset upgrades. Some of the respondents in this category indicated that they had not yet started harvesting their sisal and therefore had not benefited from sisal sales.

Agricultural assets

In assessing agricultural assets, respondents were asked to indicate what significant agricultural assets they had before and after the project. Responses were classified as follows: no change, minor, moderate, or major upgrades:

- none/no change – no increase in agricultural assets;
- minor – hand tools, up to 5 sheep/goats;
- major – up to 2 bovids (cattle/ox), tractor, trailer.

A total of eight households, or 29 per cent of the respondents, recorded major changes in agricultural assets; seven households (25 per cent) had minor increases in agricultural assets and the remaining 11 households

(39 per cent) had no changes in agricultural assets, including the respondents who indicated that they had not yet made any sisal sales: 'My income has not yet changed. I have not yet had any sales because I have not yet started harvesting my sisal'.

Access to land

Most households reported an increase of agricultural land under cultivation during the projects. The mean land area devoted to agricultural production initially was 2.3 hectares, rising to 12.4 hectares.

The yields were mostly less than 2 tonnes per hectare, much lower than potential yields, which were estimated to be 4 tonnes per hectare. One key informant, a plant protection specialist from an agricultural research institute, commented: 'Current yields for smallholder farmers are low. With good husbandry practices (good planting material, preparation of nurseries and not planting suckers), smallholder farmers can achieve the potential yield of 4 tonnes per hectare.'

Project services

A total of 25 farmers (63 per cent) indicated that they had received training, 35 (88 per cent) had received extension support, and 17 (43 per cent) received planting material from the project. Respondents were asked about the adequacy of the services provided by the project. The majority of the responses were negative: 61 per cent of the farmers rated the adequacy of the training offered by the project as inadequate; 60 per cent rated extension as inadequate; and 88 per cent rated the provision of inputs (planting material) as inadequate. No participants indicated that the level of services provision was very adequate. These findings were supported by qualitative information from the smallholder farmers:

> The project should have disseminated more knowledge on good sisal husbandry.

> More extension services should have been given focusing on increasing sisal production knowledge and skills.

> The project should have put more emphasis on technology for planting material of sisal and equipment. Currently, we are using sisal suckers instead of using pure materials.

Key informants observed:

> The project did not have sufficient funding to support training, inputs and extension to the smallholder farmers (executive director, agro-processing firm and former national project officer).

> Leaf spot disease (korogwe) is still a problem which has not been addressed and this affects the quality of sisal fibre. There is need for

continuous research on the issue (sisal plant-protection specialist, agricultural research institute).

All the respondents indicated that pests and diseases were a major constraint (particularly the korogwe leaf spot) and were inadequately addressed by the project. Access to credit was also viewed as one of the most inadequately addressed issues; 84 per cent of the farmers who responded to the question rated it inadequate: 'The project should have provided loans to assist land preparation, planting, and harvesting' and 'Credit to meet farmers' farming operations (field maintenance and purchase of inputs) should have been addressed.'

One key informant commented: 'Lack of credit is a major constraint affecting both the sisal production and productivity in the smallholder sector' (sisal plant protection specialist, agricultural research institute).

The training on post-harvest handling skills was rated only 'inadequate' or 'adequate'. Actual post-harvest handling, for storage and for the decorticator/brushing facilities, was generally perceived to be adequate as this were provided by a private agro-processing firm.

Other constraints that were highlighted during interviews were the high production and processing costs, and the competition for labour with other sectors, which contributed to the high costs.

Adequacy of marketing

Farmers did not perceive price formation to be straightforward. The smallholders produced and marketed leaf sisal but the price paid to the farmers was based on the fibre content. The processing of the leaf into fibre was undertaken by a private agro-processor which was responsible for most marketing functions. From the qualitative information collected, respondents indicated that the issues of product prices received by farmers and the costs of transport needed to be addressed. 'Farmers should be represented in the marketing and setting of prices so that they can receive a fair price'.

One key informant commented about the unrealized opportunity for smallholders to add value: 'Farmers could get a better price if they were selling fibre as opposed to the leaf. Processing costs take 60 per cent and the farmer gets the remaining 40 per cent' (official, Sisal Farmers' Association/Tanzania Sisal Board).

On transport, farmers felt that the leaf transportation service should be improved by the private agro-processor, because there were not enough vehicles: 'Leaf transportation should be improved, there are not enough vehicles/tractors to meet demand of farmers'.

Financial assets

Data on actual income gained were incomplete. However, respondents were asked whether there was income gain as a result of the project. A total of 22 farmers out of the 32 farmers who responded to the question were positive.

Five farmers indicated there was no change in their income. The reasons given were that the area under sisal was still small while input costs were high and therefore they were only able to break even. One key informant also commented: 'Major constraints that have limited production are the high input costs relating to clearing land, high cost of labour and lack of access to credit to fund sisal-farming operations' (group manager, agro-processor).

Social assets

A total of 38 of the 40 farmers interviewed indicated that they had joined the smallholder producer organizations. Some farmers in Mwelya Estate indicated that they were members of the Savings and Credit Union. One smallholder farmer commented: 'My income has increased. I can now make savings with the Savings and Credit Union.'

Conclusions

The project facilitated smallholder farmers to participate in commercially oriented sisal production activities. Sisal has increased some smallholder earnings in the drought-prone areas.

While the project design addressed partially the issue of access to land, other production aspects (inputs, technical issues) were not satisfactorily addressed. Training in sisal husbandry was inadequate, extension support was weak, and there was lack of access to credit. These factors have had an impact on the production and productivity of sisal by smallholder farmers. Securing adequate land-tenancy arrangements, as opposed to land access, was incomplete because farmers did not have title deeds to the land for use as collateral for lending. This constrained their access to credit. One key informant commented: 'A major constraint that limits farmers' participation in the market is that farmers do not have title deeds to the land and this has limited their access to credit' (official, Sisal Farmers' Association/Tanzania Sisal Board).

Human asset formation was inadequate: the project did not adequately address farmers' production and post-harvest skills. Provision of services to farmers was also deficient: there was a perceived lack of adequate transport to take produce from the farm to the processing facility. Not only was the weakness in the value chain link between producers and processor responsible for insufficient services and inputs, but inefficiencies also impaired product quality at both production and processing stages:

> The quality of fibre from smallholder producers is generally low due to poor husbandry practices. In cases where the farmers decorticate the leaves themselves with no water, the quality deteriorates further (key informant, Sisal Association of Tanzania).

The issue of targeting is also significant in this case study. Project beneficiaries were selected on a voluntary basis. There was an advertisement and anybody was free to apply. Personal data collected reflect that some participants were formally employed elsewhere and some had university degrees but there were also some low-income smallholder farmers. The question of who is targeted is important and has implications for the design of programmes aimed at alleviating poverty.

The complex design of the project included a set of different initiatives covering agricultural and industrial sectors. There were difficulties in delivery of certain components which were dependent on lagging results of others. Some components would have had better impact if they had been separate, focused projects. The major weakness of the smallholder farming system sub-component was that the issue of the productive capacity of the smallholder farmers – in terms of the productive skills of growers, extension support, and other supply-side constraints such as access to inputs – was not satisfactorily addressed.

The impact of these complexities can only be hinted at because they were not all the subject of this assessment. Nevertheless, given the geographical location of these semi-arid areas, sisal cultivation provided one of the few viable agricultural production initiatives for low-income farmers to supplement on-farm food production and generate cash incomes. Productive assets, in the form of better access to more land, were increased, but the supply of knowledge and technology to remote areas in order to leverage these productive assets was found to be problematic. It was evident that investment in human-asset skills for development of the new production systems received inadequate attention.

There was a transformational outcome for some farmers, with a degree of professionalization within agriculture as non-farmers became farmers. There were also scale effects as some smallholder farmers increased the size of their land holdings. There were income gains for most of the smallholder farmers. Nevertheless, given the long production cycle of the crop (three to four years), some farmers acknowledged that they had not yet received an income, so it was not possible to evaluate economic benefits overall. A longer timescale is necessary to assess how the project would bring financial benefits to participants, and probably also to realize the potential productivity increases from technical and technological improvements to production systems.

Social assets were boosted by the interventions. Smallholder farmers were able to organize themselves into groups and cooperative entities. This boost to social capital strengthened their position, and they gained the capacity to lobby with estate management and other entities for better returns from sisal cultivation. Integration of farmers into savings systems was a significant advantage for household risk management and financial asset building. For non-participants, there was a boost to the local economy through employment opportunities generated for sisal planting, weeding, and harvesting.

Regarding the external environment, the link to private enterprise was positive, but the monopsonistic/monopolistic nature of the relationship failed to create competition in the market for services delivery and product sales. Greater foresight concerning wider market structures and the implications of economic power for benefit distribution could have led to more equal and beneficial contractual arrangements.

Such commodity-development projects have to take into account the wider context within which interventions and activities are implemented, and the external factors which, inevitably, will condition the outcomes of different participants. Successful smallholder participation is predicated on building a viable value chain. Sisal is an example of a crop that smallholders can successfully grow, but whose viability and sustainability depends on the complex downstream processing and marketing functions.

References

FAO Committee on Commodity Problems (2013). *Potential Constraints to Smallholder Integration into the Developing Sisal Value Chain in Tanzania. CCP:HF/JU 13/3*. Rome, Food and Agriculture Organization of the United Nations.

Kimaro, D., Msanya, B.M. and Takamura, Y. (1994). Review of sisal production and research in Tanzania. *African Study Monographs* **15**(4): 227–242.

Shamte, S. (undated). *Overview of the Sisal and Henequen Industry: A Producers' Perspective. FAO Corporate Document Repository*. Rome, Food and Agriculture Organization of the United Nations. Retrieved 28 March 2017, from http://www.fao.org/docrep/004/Y1873E/y1873e05.htm.

CHAPTER 9
Strengthening the productivity and competitiveness of smallholder dairying in Zambia

Dairying is an important means of bringing economic benefits to smallholders. This study examines a dairy project in Zambia that was successfully integrated farmers into the dairy sector, with increases in productivity and value addition. An important question remaining is the targeting of the participants, who were not particularly poor households but who were more likely to be able to bear the risk of increased investment in livestock production.

Keywords: dairy, productivity, value addition, value chain, targeting, risk

Project design and implementation

Livestock projects have considerable potential to increase animal-source foods and incomes for small-scale producers (Flores-Martinez et al., 2016; Maestre et al., 2017). In particular, dairy interventions are a common and obvious way of improving incomes and livelihoods of poor people (Rawlins et al., 2014; Hoddinott et al., 2015). Jodlowski et al. (2016) recently reported a study from 2012 in Zambia which showed that goat and cow distribution by Heifer International led to improved food security through consumption of nutrient-dense foods and also to improved incomes of participants.

The potential for multiplier effects from livestock projects is important, in respect of benefits from value-adding activities such as transport and processing, increased employment opportunities, and the benefits attributable to the consumption of nutrient-intensive foodstuffs. However, livestock production is not a panacea: seasonal effects, particularly in terms of water availability and feed quality during the dry season, are often constraints to development of the livestock sector, and can lead to losses in bad years. Important potential trade-offs are the environmental consequences of poor management, particularly in emergency situations. Jodlowski et al. (2016) also pointed out that the increased labour requirement for dairy management may fall disproportionately on women, and that it is a risky enterprise, not necessarily feasible for the poorest households.

Project context

The project analysed here was implemented in Lesotho and Zambia over a four year period up to 2012, and was aimed at improving the productivity and marketing position of smallholder dairy cooperatives. Resolving the problem of dry-season feedstuffs was given particular attention. In Zambia, the implementing agency was Golden Valley Agricultural Research Trust and the project targeted 730 smallholder dairy-farmer households in the five districts of Kaloma, Choma, Monze, Magoye, and Mapepe. The three main project components were:

- *Component 1: Production intensification.* Promote improved and innovative livestock feeding technologies for local production and conservation of protein-rich feedstock focusing on the dry season (May–June).
- *Component 2: Value and quality addition.* Enhance milk quality and hygiene to reduce wastage and increase shelf life and safety of the milk.
- *Component 3: Project extension.* Improve information dissemination through exchange visits, TV and radio programmes, field days, and production of leaflets and pamphlets.

The direct beneficiaries of the project were the smallholder dairy farmers. Secondary beneficiaries were considered to be the milk processors and also consumers who benefited from increased local milk supplies and better-quality milk.

Thirty-nine farmers (25 male and 14 female) from seven districts, mainly in the southern provinces of Zambia, were interviewed. Respondent numbers are shown in Table 9.1.

The majority of the farmers interviewed were male heads of families (31 of the 39 households were headed by men). Most of the farmers were over 50 years of age and the level of education was very high; most had completed secondary education and had also attained a tertiary qualification in the form of a certificate or a diploma. Further, the majority of these farmers had lived in these districts for more than 10 years (Table 9.2).

The mean distances of the respondents from the main road and the milk-collection point were both just over 5 kilometres, a relatively small distance for transport given adequate facilities, and the mean distance to the major service centre was about 18 kilometres.

Table 9.1 Farmer geographical distribution

Districts	Number of beneficiaries interviewed
Kaloma	4
Chilanga	9
Kafue	1
Choma	4
Southern Mazabuka	2
Monze	10
Mazabuka	9
Total	39

Table 9.2 Personal data and project participation

Age category		Level of education		Tertiary qualifications		Date of organization membership		Years living in the area	
25–30	12	Primary	7	None	13	Pre-2000	2	<10	3
30–50	1	High school	32	Certificate	17	2000–07	35	10–30	21
>50	25			Diploma	8	Post-2008	1	30–50	8
				Degree	1			>50	6

Table 9.3 Changes in household assets and land utilization

Assets	Before project	After project
Bicycle	20	22
Radio	30	35
Television	23	32
Land utilization (ha)		
Mean	10.2	13.9
Minimum	0.0	1.0
Maximum	100	100
Standard deviation	18.9	20.3

Findings

Household and agricultural assets

Respondents were asked to indicate significant household assets owned before and after the project. Asset accumulation was indicated by increases in ownership of bicycles, radio, and television sets after the project (Table 9.3).

Land utilization by respondents also increased from a mean of 10.2 hectares to 13.9 hectares, attributed by farmers to the need for more land to grow feed for the dairy cows. There was considerable variation among respondents.

Project services

A total of 37 of the sample of 39 farmers indicated having received training, 33 received extension support, 32 received feed-planting material, and 29 received information from the project. The majority of respondents considered the services provision to be adequate and only 19 per cent considered the services inadequate. Inputs and information provision scored very low (Table 9.4).

Responses concerning the issues identified by participants as constraints to the production of milk in the various districts varied. For example only six farmers considered the distributed inputs to be inadequate, compared with 20 who considered that the provision of credit was inadequate. A small

Table 9.4 Adequacy of project services provision

	Training	Extension	Inputs: planting material	Information
Inadequate	4	14	10	2
Adequate	22	15	12	11
Very adequate	10	7	4	2
No response	2	3	13	24

majority considered that the constraints affecting production were addressed adequately, against 30 per cent who felt that these issues were inadequately addressed.

Post-harvest handling was generally perceived to be adequate across the three attributes of skills, information, and storage. The provision of generators by the project at milk-collection centres helped to cut milk storage losses significantly, ensuring a consistent supply of fresh milk to the market.

With respect to a range of factors, the marketing of milk was considered to have been adequately addressed. These factors covered milk quality and consistency of supply, marketing information, supply-chain linkages, collection, and transport. These improvements in the marketing of the milk were attributed by farmers to collective selling, which ensured that all their milk was purchased, unlike the situation previously, when they had marketed the milk individually to the public and through local kiosks.

Improvement in market participation

A raft of indicators was used to evaluate the enhanced market participation and productive capacity, these included farmers' ability to supply more milk (measured by an increase in the number of dairy cows), the survival rate of calves, milk output per cow, and the total amount of milk produced per day. This information was further supported by interviews with major milk-processing companies in Zambia.

The farm-gate price for raw milk increased from US$0.30–0.35 per litre in 2007 to US$0.50–0.60 in 2012. Moreover, milk intakes by the processors from the smallholder dairy/milk-collection centres increased from 8 per cent in 2007 to 20 per cent for one of the buyers, Parmalat. Dairy King was the sole buyer of milk from Mapepe Dairy Cooperative and the intake increased from 30 per cent in 2007 to 70 per cent in 2012. According to officials from Parmalat and Dairy King, the quality of milk from the smallholder dairy farmers/milk-collection centres improved from Grade C to Grades A and B. Industrial buyers noted that quality was enhanced by the provision of milk cooler tanks and this was sustained by the provision of generators by the project which dramatically reduced milk spoilage: by 90–100 per cent compared to before the project when all milk went to waste because of electricity power failure/load shedding. This is important evidence of value addition through improved milk quality and processing which resulted from both technical advances and effective investment in value chain linkages. The benefits of this were shared by the primary producers.

The increased price range and the enhanced value chain linkages were seen as sustainable, resulting in net returns to farmers of more than US$0.25 per litre, making smallholder dairy a sustainable source of daily disposable income. It must be noted that the improved participation of dairy farmers cannot solely be attributed to this project, as a number of projects had been implemented in the past and some even overlapped with this project.

Viability of milk collection centres

There was an improvement in utilization of installed milk-cooling capacity from 22 per cent in 2007, when the project had just started, to 68 per cent in the dry season and 100 per cent in the peak period in 2012. This was also reflected in increased incomes generated at the milk-collection centres.

Changes in the number of dairy cows

Of the 39 respondents, nine had no dairy cows before the project. Herd sizes increased, and 32 of the respondents attributed the increase in the number of their dairy cows to the project (Table 9.5).

Productivity measures

An increase in the survival rate of calves was attributed by most respondents to the activities of the project, which helped to enhance animal husbandry.

Milk production per cow rose significantly from a mean of 4.1 litres to 8.4 litres per day. With increased cow numbers and increased productivity, participating farmers saw total milk output per day rise from an average of 29.8 litres to 81.3 litres.

Milk losses were eliminated through the installation of generators by the project: the mean loss of 30 per cent fell to zero per cent.

Table 9.5 Changes in dairy herd size

Herd size	Before project	After project	Percentage change
0	9	0	−100
1 to 5	18	6	−66.6
6 to 10	5	11	120
11 to 49	4	19	375
Greater than 50	1	1	0

Conclusions

The project facilitated farmers' penetration into the commercial milk-production chain in Zambia. The productive capacity of the dairy farmers improved significantly and post-harvest handling also improved significantly and this is reflected by a dramatic reduction in milk losses and also an improvement in milk quality. The collective system helped to boost the marketing of milk since, as most of the farmers pointed out, this arrangement has helped provide farmers with an assured milk outlet. Overall, the project components had a positive impact, although there were weaknesses, and not all the benefits could be attributed to the specific intervention. Delivery of infrastructure and logistics components appears to have been more effective than that of the softer human-capital skills and technical and advisory services to the farmers themselves. We are unable to comment on the feedstuff component directly, but the evidence of increased animal reproduction and survival rates, and increased milk productivity suggests that this important component was not a constraint to the success of the project.

The favourable project environment was attributable to the involvement of government, the private sector, and donors, which created a positive and comprehensive institutional environment that complemented the project and made it possible for this intervention to yield positive results. Of particular significance was the creation of efficient value-adding chain linkages between the key actors, and the creation of employment opportunities within processing links in the chain.

One curious and unsatisfactory feature of the project remains to be discussed. The selection of the participants for this project was not clearly defined and articulated. The target group was meant to be smallholder farmers; however, without a proper or agreed definition of a smallholder farmer, it is difficult to tell whether some of the project beneficiaries were supposed to be included in this project. The profile of beneficiaries was not typical of smallholder farmers needing improved market access: of the sample of farmers interviewed, more than half were over the age of 50, and many had tertiary education qualifications. Doubts persist about the targeting and additionality of the project.

Targeting of agricultural projects designed to benefit poor peoples is an issue of general importance. It becomes more complicated in the case of livestock projects. As noted by Jodlowski et al. (2016), livestock-development projects are capital-intensive and relatively high risk, as well as potentially high reward in terms of nutritional, health, and economic benefits – to say nothing of the benefits arising from integrated livestock–agriculture systems, and the savings and insurance function of livestock. There is a danger of impoverishment when a project goes wrong – such as the loss of a market in the case of Ethiopia – and it can be argued that livestock production is too risky for the most vulnerable rural people because of multiple threats, not least from costly losses through disease, theft, and adverse climatic conditions.

Summarizing: there were significant improvements to the productive capacity of the dairy farmers, with better handling systems that brought about big reductions in milk losses and improvements in milk quality, and valuable price premiums. The value chain was strengthened through effective collective organization which facilitated processing and marketing. In terms of physical capital, significant investment was evidently attributable to the project. Investment in farmers' human assets was more limited.

References

Flores-Martinez, A., Zanello, G., Shankar, B. and Poole, N. (2016). Reducing anemia prevalence in Afghanistan: socioeconomic correlates and the particular role of agricultural assets. *PLoS One* **11**(6): e0156878 <https://doi.org/10.1371/journal.pone.0156878>.

Hoddinott, J., Headey, D. and Dereje, M. (2015). Cows, missing milk markets, and nutrition in rural Ethiopia. *Journal of Development Studies* **51**(8): 958–975 <http://dx.doi.org/10.1080/00220388.2015.1018903>.

Jodlowski, M., Winter-Nelson, A., Baylis, K. and Goldsmith, P.D. (2016). Milk in the data: food security impacts from a livestock field experiment in Zambia. *World Development* **77**: 99–114 <https://doi.org/10.1016/j.worlddev.2015.08.009>.

Maestre, M., Poole, N. and Henson, S. (2017). Assessing food value chain pathways, linkages and impacts for better nutrition of vulnerable groups. *Food Policy* **68**: 31–39 <http://dx.doi.org/10.1016/j.foodpol.2016.12.007>.

Rawlins, R., Pimkina, S., Barrett, C.B., Pedersen, S. and Wydick, B. (2014). Got milk? The impact of Heifer International's livestock donation programs in Rwanda on nutritional outcomes. *Food Policy* **44**: 202–213 <https://doi.org/10.1016/j.foodpol.2013.12.003>.

CHAPTER 10
Assessing the impact of projects on smallholder participation in agricultural markets: Synthesis and conclusions

In this chapter we present a synthesis and comparative analysis of the studies presented in Chapters 6–9 based on a regression study of the data for all four cases. Overall conclusions are presented.

Keywords: regression analysis, asset changes, information and training, capacity development, project impacts, market participation, threshold, targeting, project design and implementation, sustainability

Recapitulating the theory of change: expected relationships

In the previous chapters we used sets of descriptive statistics to describe and compare variables that influenced smallholder farmers' access to markets. This chapter gives an account of a quantitative analysis based on a regression technique to test factors affecting improvements in market access. These factors refer to the five asset indicators developed in Chapter 5 and illustrated in Figure 5.1. Improved market access was defined as an increase in the price received for the produce as well as improvements in terms of quality and quantity. All this information was collected from the field surveys. The dependent variable was the natural log of the probability of improved access to market divided by the probability of no improvements (1-P). In this case, a maximum likelihood estimation was preferred as it yields consistent, efficient, and asymptotically normal estimators (for further information on the quantitative methodology, see Amrouk et al. 2013).

With respect to the expected signs of the independent variables, it was hypothesized that positive changes in household assets as a result of project activities are positively related to the probability of improved market access of the participating smallholders. Ownership of *household assets* such as a radio, bicycle, and motorbikes increase farmers' access to information and their likelihood of participating in markets. *Extension services* delivered under the projects covered a wide range of activities ranging from crop husbandry to dissemination of information on input supplies and product markets. Thus, access to extension services was expected to positively influence the probability of participation in markets. Similarly, smallholders who had access to *credit* were considered more likely to acquire factor inputs and thus more likely to participate in output markets. For example, the establishment of a revolving loan fund was expected to be positively associated with the probability that smallholders would improve their market participation. Access to *productive assets* such as planters, cultivators, and technology should enhance the ability of farmers to generate marketable production surpluses. Thus, a build-up of productive assets through the projects would increase the probability of successful market participation. Likewise, project activities which developed *skills* in cropping, marketing, and farm management would be likely to create enabling conditions for smallholders to enter markets. Increased *skills*, therefore, were expected to relate positively to the dependent variable. Finally, farm and household characteristics, such as *education, wealth, age, farm size*, and *location* were assumed to influence positively the probability of improved market access. The *location* variable captured regions with higher potential as a result of, for example, more suitable soil types, climate, and local market structure.

A number of lessons can be drawn from the determinants of successes and failures of the projects. The next section discusses comparative findings based on the qualitative and quantitative analyses.

Lessons learned

Overall results

Overall, project activities had a positive impact on market participation. In order to assess empirically how changes in asset indicators influenced market participation, a logit regression analysis was applied to the entire set of survey data. The selected independent variables covered the asset indicators from the livelihood assessment approach as described in Table 10.1. This shows summary and comparative results for all cases broken down into the five livelihood assets: natural, human, social, physical, and financial. Household and farm characteristics contributing significantly to the model were added on the basis of the resulting goodness of fit measured by the log likelihood chi-square. Also, variables with multicollinearity were identified and dropped from the model.

The variables *education, wealth, farm size, district*, and *age of the household* were found to improve the explanatory power of the model. The resulting odds ratios and standard errors are shown in Table 10.2. The likelihood ratio chi-square of 46.4 with a *p*-value less than 0.05 shows that the resulting model as a whole fits significantly better than a model with no independent variables. This means that the explanatory variables included contribute jointly towards explaining smallholders' market participation patterns. *Access to extension services and training, expansion in private productive agricultural assets, access to credit, district, farm size, age of the household*, and *wealth* had significant influence on market access. All these variables were significant at the 10 per cent level, except for the variables *access to credit* and *district*, which were significant at the 5 percent level.

Table 10.1 Impact of projects

Asset indicators	Ethiopia	Peru	Tanzania	Zambia
Natural	Positive. Land devoted to green beans production expanded	Positive changes. The scale of jatropha production increased	Access to land was increased and farmers adopted a new hybrid sisal variety	82% of respondents attributed increases in dairy cows to project
Human	Positive changes. Skills acquired in green-bean production and quality control. New market exploited	Positive changes. New skills acquired in crop husbandry including pruning, use of manure, bioinsecticides, use of bees	Limited changes in human asset building as farmers had limited production skills and extension support	Positive changes. Farmers improved skills in milk production and post-harvest handling
Social	None. No access to the market beyond the project or sustainable market linkages	Positive changes. Farmers formed agricultural associations, social interaction among farmers improved	Positive changes. Farmers able to organize themselves into cooperative groups and unions	Positive changes. Farmers organized into cooperatives to sell milk in bulk for processing
Physical	Increases in agricultural assets and household assets for most households	Increases in physical and agricultural assets; investments in housing	Increases in agricultural assets and household assets	Increases in physical assets for most households
Financial	Positive changes in incomes were noted	Income gains for some farmers. Most were expected to make higher incomes after sixth year of production	There was income gain for most smallholder farmers	There was income gain for most smallholder farmers

A positive and significant relationship ($p = 0.066$) was found between improvements in market access and the provision of *extension services and training*. A shift from inadequate access to extension to adequate access to extension increased the probability of improvements in market access from 0.280 to 0.865, which corresponds to a 58.5 percent increase (Table 10.3). This result underlines the importance of extension services and training for dissemination of knowledge, technology such as new crop varieties, and improved farm practices.

Similarly, a positive and significant relationship was found between changes in *agricultural assets* associated with the project and market participation. The probability of improvements in market access increased by

Table 10.2 Influence of asset indicators on market participation

Acc to market	Odds ratio	Std err	z	p<z	95% confidence interval	
Hous_asset	0.985	0.064	−0.22	0.827	0.867	1.120
Acc_ext	1.018	0.010	1.84	0.066	0.998	1.038
Acc_credit	5.175	3.967	2.14	0.032	1.151	23.254
Ag_asset	1.090	0.050	1.87	0.061	0.995	1.193
Skills	1.850	1.608	0.71	0.479	0.337	10.161
Farm_size	0.892	0.055	−1.82	0.069	0.789	1.008
Age	0.175	0.112	−2.72	0.006	0.050	0.614
District	1.506	0.212	2.91	0.004	1.143	1.985
Education	0.992	0.293	−0.02	0.981	0.556	1.771
Wealth	10.887	15.270	1.7	0.089	0.696	170.136

Note: Logistic regression: Log likelihood = −34.16435; Number of obs = 84; LR chi^2(7) = 46.4; Prob > chi^2 = 0.000; Pseudo R^2 = 0.4044.
The z-value or z-statistic is the regression coefficient divided by its standard error. A z-value greater than 2 suggests that the statistic is significant.
A small p-value, commonly taken to be less than 0.05, also suggests that the relationship between variables is significant.

Table 10.3 Probability changes in market access for changes in explanatory variables

Change in variable	Probability	Change in probability of market access
Access to extension		
Inadequate to adequate	0.280–0.865	+0.585
Access to credit		
Inadequate to adequate	0.348–0.924	+0.576
Wealth		
Poor to middle-income	0.121–0.729	+0.601

a factor of 1.1 for smallholders who built up *agricultural assets*, such as livestock and planters, as a result of their involvement in the projects, in comparison with those who did not manage to build up productive assets. In many cases, livestock sales provided smallholders with the liquidity necessary to purchase fertilizer, high-yielding varieties, and other technologies to increase the marketable surplus. The evidence shows that projects which build domestic assets can generate significant returns to smallholders.

The variable *credit* was found to be significant ($p = 0.032$) and had the expected positive sign, underlining the stimulating effect of project activities that include the provision of financial services. An example was a revolving loan fund for smallholders established in the case of the project in Ethiopia. The analysis of the partial effect of the variable *credit* showed that a shift from having inadequate access to credit to adequate access to credit increased the probability of improvements in market access from 0.348 to 0.924, or by 57.6 percent (Table 10.3).

The result from the logit regression also showed that the likelihood of an improvement in market access increased with the geographic location, or *district*. Smallholders located in districts with a tradition of commercialization were more likely to engage in market transactions themselves. The variable *district* was found to be positive and significant ($p = 0.004$) (Table 10.2), with the probability of improvements in market access increasing by a factor of 1.51 for smallholders located in districts with prevailing favourable agroecological conditions, socio-economic structure, and/or other related agriculture and local factors (Table 10.3). A positive and significant ($p = 0.089$) relationship was also found between market access and *wealth* of the smallholders (Table 10.2). Wealthy and better-endowed smallholders were more likely to benefit from project activities and further raise their level of market integration. In fact, a shift from being a poor smallholder to a middle-income smallholder increased the probability of improvements in market access from 0.121 to 0.729, or by 60.1 percent (Table 10.3).

The variable *farm size* was found to be significant and have a negative sign. This means that smallholders with large farms felt that project interventions did not enable them to improve their market participation. This unexpected inverse relationship may have been due to diseconomies of scale: for example, gains in land productivity and/or market sales were not large enough to offset the costs associated with the increase in production.

Finally, the variable *age* of the household head was significant ($p = 0.006$) and negatively related to market participation, suggesting project interventions tended to benefit younger farmers more than older farmers (Table 10.2). This is probably due to the ability of younger farmers to assimilate new technologies and innovative business practices more efficiently, enabling them to overcome fixed market-access costs such as transaction costs. Other variables such as *change in household assets*, *skills*, and *education* were not significant at the 10 per cent level. However, their inclusion was found to improve the logit model, suggesting that their association with other independent variables contributes jointly to explaining smallholder market participation patterns.

Clearly, there were synergies among the different market-participation indicators due to synergies and spillovers that exist across the various dimensions of market participation: for example, stimulating access to credit is likely to lead to greater access to inputs, improved technology adoption, higher production, and better market linkages. Delivery of comprehensive services is necessary to take advantage of this relationship. Considering the importance of training and extension, a key challenge is to maintain funding to extension units beyond the lifetime of the project, so that gains in market access are sustained and further developed. Where funding is inadequate for comprehensive service provision, and project design does not recognize time constraints, the challenge for policy and project interventions is to adequately select and prioritize the constraints that appear to be the most limiting.

Impacts of projects on smallholder livelihoods

Project activities did have relatively significant positive effects on smallholder livelihoods. In Ethiopia, it was not possible to compare income changes in the absence of a baseline, but project beneficiaries acknowledged that revenues and livelihoods improved, a situation they associated with the implementation of the project. Returns from the survey also showed that farmers were able to increase the scale of agricultural production overall, while analysis of assets suggested a number of benefits through reinvestment of income in minor household assets, improvements in housing infrastructure, and acquisition of new agricultural assets. There was also capacity building among farmers: significant skills in production and post-harvest handling were gained. The construction of two pack-houses was a major investment for the cooperatives, enabling smallholders to process value added products and local communities to benefit through the creation of new employment at both the farm and processing levels.

In Peru, smallholders learned new techniques for the production of jatropha, including pruning, use of organic manure, bioinsecticide use, and use of bees that help to pollinate the crop. Beekeeping became a source of additional income and improved livelihoods for farmers who consumed and sold the honey in the market. The project also helped the development of a bioinsecticide made from jatropha oil. This organic product has fungicidal properties and can be used as an insect repellent. Given the opportunities created by the expansion of organic agriculture, smallholders participating in the project were able to generate extra income through the sale of organic fertilizers. Farmers were also able to invest in household assets and make improvements to housing infrastructure as well as acquiring new agricultural assets. Finally, the demand for agricultural labour increased in the community as a result of the project activities.

In the semi-arid environment of Tanzania, sisal cultivation provided low-income farmers with an alternative agricultural activity that generated a sustainable income to supplement food production. In addition, access to land was increased, allowing some farmers to expand their agricultural holdings and the overall scale of production. Results showed that sisal-project activities enabled an increase in productive assets and led to income gains for most participants. Smallholders were also able to organize themselves into groups and cooperatives. This strengthened their position to lobby for better input and sisal prices. At the community level, demand for labour rose in response to sisal planting, cultivation, and harvesting operations.

The project in Zambia showed that strengthening market participation of dairy farmers could be achieved through the diffusion of innovative livestock-feeding technologies and the conservation of protein-rich feedstock with a focus on the dry seasons (May–June). Milk production per cow rose significantly from a mean of 4.1 litres to 8.4 litres per cow per day, a 104 per cent increase. About 82 per cent of the respondents attributed this change to the project. Milk quality increased, allowing for longer shelf life and improved product safety. From a mean loss of 30 per cent, milk losses were eliminated as a result of the installation of generators for cooling milk funded by the project. At the community level, there was an increase in employment prompted by increased production and processing of milk.

In the four case studies, improved access to privately held assets and technology, along with knowledge of new production and post-harvesting skills, helped raise overall productivity and access to markets. As shown in Tables 10.4 and 10.5, both household and agricultural assets increased. In the latter case, the number of assets was higher following the implementation of the project activities, with the difference statistically significant at the 5 per cent level (Table 10.4). The increase reported in household assets due to participants' involvement in the project was statistically significant at the 5 per cent level (Table 10.5). About 69 per cent of the respondents believed that transfer of technology was appropriately provided during the project.

One of the key initiatives to expand privately held assets by smallholders was to improve access to financial services such as credit and savings. In the case

Table 10.4 Paired t-test on access to agricultural assets

Variable	obs	Mean	Std err.	Std dev.	[95% Conf. interval]	
agAsset2	132	10.36364	1.823944	20.95552	6.75544	13.97183
agAsset1	132	6.621212	1.299279	20.95552	4.050929	9.191495
diff	132	3.742424	1.131153	12.99596	1.504734	5.980115
mean(diff) = mean(agAsset2 − agAsset1)				$t = 3.3085$		
Ho: mean(diff) = 0				Degrees of freedom = 131		
Ha: mean(diff) < 0			Ha: mean(diff) = 0		Ha: mean(diff) > 0	
Pr(T < t) = 0.9994			Pr(T > t) = 0.0012		Pr(T > t) = 0.0006	

Note: agAsset2: Frequency of agricultural assets after project implementation; agAsset1: frequency of agricultural assets prior to project implementation

Table 10.5 Paired t-test on access to household assets

Variable	obs	Mean	Std err.	Std dev.	[95% Conf. interval]	
hAsset2	132	2.621212	0.371543	4.268704	1.88621	3.356213
hAsset1	132	1.962121	0.1715361	1.970799	1.62278	1.36817
diff	132	0.6590909	0.3584397	4.118159	−0.0499883	5.980115
mean(diff) = mean(hAsset2 − hAsset1)				$t = 1.8338$		
Ho: mean(diff) = 0				Degrees of freedom =131		
Ha: mean(diff) < 0			Ha: mean(diff) = 0		Ha: mean(diff) > 0	
Pr(T < t) = 0.9659			Pr(T > t) = 0.0682		Pr(T > t) = 0.0341	

Note: hAsset2: Frequency of household assets after project implementation; hAsset1: frequency of household assets prior to project implementation of the project in Ethiopia, the establishment of a revolving loan fund enabled smallholders to purchase inputs necessary for profitable commercial green-bean production. The vast majority of respondents felt that provision of credit was adequate. However, when aggregating the results across all four projects, 50 per cent of the surveyed smallholders felt that the issue of lack of credit was not adequately addressed through project activities.

On the other hand, access to extension services and technology was considered adequate (Tables 10.6 and 10.7). On aggregate, improved access to technology was associated with better market integration for 67 per cent of the survey respondents. The impact of technology adoption can be assessed by looking at changes in productivity levels. For example, Table 10.8 compares yields for green beans before and after project implementation in Ethiopia. It shows that yields were relatively higher following the project, and the difference between yield levels was statistically significant at the 5 per cent level. Similarly, in Zambia, milk production per cow increased by more than 100 per cent. About 82 per cent of the respondents attributed this change to the project (Table 10.9).

Shortcomings in project design and implementation

As discussed in the previous section, returns from the field survey illustrated the positive effects of the project activities on smallholders' market participation, particularly with reference to access to private productive assets. However, the survey results also evidenced the need to focus on three main areas in the formulation and implementation of market participation projects.

First, the selection of deserving project beneficiaries has to be undertaken on the basis of a systematic approach. As shown in Table 10.10, 87.7 percent of the targeted smallholders considered themselves as middle income, while only

Table 10.6 Access to extension services through projects

Ext. support	Frequency	Percentage	Cumulative
Inadequate	30	29.4	29.4
Adequate	48	47.1	76.5
No response	24	23.5	100
Total	102	100	

Table 10.7 Diffusion of technology through projects

Technology	Frequency	Percentage	Cumulative
Inadequate	8	20.5	20.5
Adequate	22	56.4	77
No response	9	23.1	100
Total	39	100	

Table 10.8 Paired t-test on green beans yields, before and after project

Variable	obs	Mean	Std err.	Std dev.	[95% Conf. interval]	
a91	69	3.175362	0.3698022	3.071808	2.437434	3.913291
a90	69	0.115942	0.115942	0.9630868	−.1154167	0.3473007
diff	69	3.05942	0.3465598	2.878742	2.367871	3.75097
mean(diff) = mean(a91 − a92)				t = 8.828		
Ho: mean(diff) = 0				Degrees of freedom = 68		
Ha: mean(diff) < 0		Ha: mean(diff) = 0			Ha: mean(diff) > 0	
Pr(T < t) = 1.0000		Pr(T > t) = 0.0000			Pr(T > t) = 0.0000	

Note: a90: green bean yields (tonnes/ha) before project; a91: green bean yields (tonnes/ha) after project

Table 10.9 Percentage changes in milk output per cow (litres), before and after project

	Before the project	After the project	Percentage change
Mean	4.1	8.4	104
Minimum	0.2	1.4	600
Maximum	14	20	42

Table 10.10 Smallholders by wealth distribution (self-reported)

Wealth	Frequency	Percentage	Cumulative
Very poor	3	2.3	2.3
Poor	11	8.5	10.8
Middle income	114	87.7	98.5
Rich	2	1.5	100.00
Total	130	100.00	

8.5 per cent and 2.3 per cent considered themselves as poor and very poor, respectively. This implies that project activities effectively targeted the better-endowed smallholders. As noted in the conclusions, this feature of targeting may – or may not – be an intentional objective of smallholder market-access projects.

Table 10.11 illustrates this targeting phenomenon, showing that 112 out of 128 (87.5 per cent) of smallholders were middle-income households with an average landholding of 5.3 hectares, which is 208 per cent more than households classified as poor. Middle-income smallholders also had a higher level of education (Table 10.12) and reported, on average, having more household assets prior to the project than poor households.

Second, the field survey revealed weaknesses in the design and implementation of the marketing component of the projects, which should have included activities specifically designed to build sustainable value chain linkages. Participating smallholders generally felt that their skills in marketing

Table 10.11 Smallholders' landholding by wealth distribution

Wealth	Mean (ha)	Std. Dev.	Frequency
Very poor	0	0	3
Poor	1.7	2.15	11
Middle income	5.3	11.47	112
Rich	1	1.41	2
Total	4.8	10.83	128

Table 10.12 Smallholders' level of education by wealth distribution

Wealth	Mean	Std. Dev.	Frequency
Very poor	3	1.73	3
Poor	2.6	0.51	11
Middle income	4.6	12.71	114
Rich	3	0	2
Total	4.3	11.91	130

Note: Education level 1: None; 2: Primary; 3: Secondary; 4: High school; 5: Diploma; 6: University

and negotiation needed to be improved, and – unsurprisingly – that prices received for their produce could have been higher. Where product markets were not characterized by a workable level of competition, farmers will have been disadvantaged. Thus, in the cases of Ethiopia, Peru, and Tanzania, lack of better prices was associated with limited market outlets; in these projects, there was only one buyer. The failure to build sustainable supply linkages was pronounced in Ethiopia. Little attention was paid to capacity building at the cooperative level and there were no enduring supply chain linkages or business networks established between producers, cooperatives, and viable export partners. For some of the smallholders, the short-term benefits of exporting green beans were outweighed by losses incurred following the decision by the exporter associated with the project to halt its green-bean operations, turning a manageable business risk into a major implementation failure. Clearly, the loss of the export channel meant that none of the direct benefits generated at the beginning of the project was sustained. Farmers who invested in specific assets for green-bean production eventually found themselves with limited or no returns on these new assets, which was a highly undesirable outcome.

Third, project agreements should have included a detailed breakdown and an assessment of the counterpart contribution, so that risks were well-balanced between the stakeholders. One specific example is again the case of Ethiopia where the required counterpart contribution, which would have locked in the exporter, was not delivered – another major implementation failure. One of the options in the case of Ethiopia would have been to expand the commodity portfolio addressed by the project along with setting up contractual agreements with more than one green-bean exporting company, as was envisaged in the project design.

Participation of smallholders in commercially oriented production and market activities

Project activities aimed at improving smallholders' market-participation capacity led to positive outcomes for farmer-beneficiaries and to local communities, who were indirect beneficiaries. Building market-participation capacity, by, for example, improving farming skills and private productive assets, is a requirement for improved market access. The evidence from the four country studies showed that smallholders' access to markets improves following the purchase of specific private assets and implementation of skills enhancement activities. In Tanzania, the release of land deeds by the government to the Tanzania Sisal Board (TSB), for allocation to smallholders to engage in sisal cultivation and supply the processing mills, was an institutional change as well as asset-building and ensured smallholder integration into the sisal value chain. Similarly, for the vegetable export development project in Ethiopia, local authorities allocated land for the construction of two privately managed pack-houses, which was a determinant in enhancing smallholders' access to markets.

However, sustaining market participation means that consideration must also be given to increasing the 'ease of doing agribusiness' (World Bank, 2017) by reducing unnecessary regulations, reinforcing market-related institutions such as the rule of law, and strengthening infrastructure such as roads, electricity, and physical marketplaces. These elements have a significant influence on farmers' decisions to participate in markets and in ensuring the success of projects. They also underpin the key role of governments as a provider of public goods. The responsibility of the public sector in commodity-specific projects must be clearly defined with identified roles and expectations. Where the public sector provides start-up support to projects in terms of financial and human resources and specific services in the fields of research, infrastructure, extension, and training and capacity building, an appropriate time frame must be adopted in order to achieve sustainability.

Institutional support, in relation to the establishment and strengthening of organizations such as cooperatives and producers' organizations, appears to be a very important component of successful projects, and is another element that often takes longer than anticipated. In the case of Zambia, Tanzania, and Peru, farmers pointed out that collective marketing secured remunerative market outlets for farmers. Smallholders' organizations are more likely to achieve economies of scale in production and purchasing of inputs, facilitating their integration into commercial farming.

The selection of specific market channels may constrain smallholders to a particular value chain as the case study in Ethiopia and, to a certain extent, in Peru showed which tends to shift a significant share of market risk on to smallholders themselves. The use of formal written contracts between producers and buyers is likely to reduce transaction-cost risks inherent in such asset-specific projects.

Capacity development must also encompass training of smallholders (and extension agents) on such issues as evaluation of market conditions and outlook, and negotiations on pricing and payment terms. This is an area in which survey respondents felt that more needed to be done to better equip them to face the challenges of operating in the market in a sustainable manner. Such requests were formulated in all four country case studies.

Recommendations for selecting project-participating smallholders

The basis, or process, for selecting project-participating smallholders is critical and needs to be clearly formulated. The first step is to agree on a set of criteria that define a smallholder. The definition may vary from project to project and also within a project itself depending on the country context. It is also equally important to identify a methodology which enables the identification and targeting of smallholders. Evidence from the case studies showed that the selection of smallholder participants was not sufficiently defined and articulated. Targeting of smallholder farmers was more successful in Ethiopia, where farmer cooperatives were involved in identifying and selecting project participants from within their own membership. In Tanzania, where the selection of beneficiaries was on a voluntary basis, there were project participants who were not necessarily in need of support. A similar situation was reported in the case of Zambia and Peru. Clearly, in many instances, targeting is fraught with ethical issues that are not frequently discussed openly. Targeting involves explicit biases. Many projects, in effect, target their interventions at the producers most likely to respond, which creates a productivity bias. Often, targeting is based on a minimum or threshold concept of asset endowments that project participants must meet:

- level of human capital, e.g. education;
- financial capital, e.g. loan collateral;
- physical and natural assets, e.g. minimum scale of production;
- social capital, e.g. cooperative membership, geographical proximity to markets, gender;
- psychological capital, e.g. perceived propensity towards market orientation and innovation.

Such an approach is likely to generate the best economic returns to the intervention. A consequence is that the most marginal and poorest rural people are likely to be excluded. However, they may benefit through second-round effects, such as the demand for labour, as was the case in Ethiopia with the establishment of the pack-houses.

The threshold concept implies the exclusion of the poorest smallholders within a community, or region, and is likely to give rise to increasing inequality.

Where specific localities are targeted, increasing inequality between regions is also a likely consequence. Adopting a productivity bias is therefore an effective approach for boosting the local agricultural economy but not an approach conducive to reducing poverty among the poorest.

An alternative method is to target the poorest, as is the case in microfinance projects. This is a poverty bias, based on common methodologies such as wealth ranking and more-or-less readily available household-income data or per-capita conceptions of wealth/poverty. The capacity of the poorest people to respond to opportunities may indeed be constrained and give lower returns to interventions. However, even small impacts on poverty may result in significant benefits to participants. Other potential and combined approaches involve targeting women, single-headed households, younger people, or collective organizations. The targeted population is, therefore, a choice variable. The decisions and the consequences should be clearly articulated. Project design must involve explicit discussion and negotiation between donors and host-government bodies, implementing agencies and potential beneficiary populations.

Project design must ensure that the objectives of the intervention are consistent with the targeting, and the objectives should preferably be limited in number and scope. Thinking in terms of livelihood assets and outcomes is a helpful way of ensuring that objectives, targeting, types of intervention, and expected outcomes are coherent. Thus, investment in human, social, financial, and physical assets (and maybe natural assets by way of land tenure measures and reforms, and investments in ecological sustainability) are proximate objectives for economic development and poverty reduction.

Commitment of stakeholders to fulfil their obligations under the project

Working through existing institutional structures, such as extension services and cooperative unions with which farmers are already familiar, is important in securing beneficiaries' buy-in to the project. This strategy was particularly successful in Ethiopia and Zambia. In Ethiopia, the project worked through the existing extension services and farmer cooperatives. In Zambia, the project built on earlier initiatives for smallholder farmers: the dairy cooperatives and the milk-collection centres. Hence the local institutional and organizational context is important. Stakeholder mapping and analysis is a necessary part of the project-design process in order to, identify and make explicit the conflicting, competitive, and cooperative interrelationships.

There is also a need for strong support and commitment by stakeholders to fulfil their obligations under the project. Contractual responsibilities should be clearly negotiated and specified. Where required, counterpart contributions must be quantified and fair. The case study from Ethiopia demonstrates that contractual commitments by stakeholders must be given the utmost attention to ensure that project results extend beyond the life of the project.

Capacity development

Participatory approaches and training are necessary to increase smallholder productive capacity and these were successfully implemented in the Zambia and Ethiopia projects. At the institutional level, there is a need to enhance the capacity of national implementing institutions. In Ethiopia, there was no capacity development at the cooperative level or sustainable chain linkages between producers, cooperatives, and export partners. In Tanzania, there was no provision made for the development of commercial-scale distribution of improved sisal planting material.

Where smallholder farmers are developing associations, it is essential to provide training to the associations, particularly in price formation, negotiation, financial management, and other business skills, to help them better fulfil their roles. Returns from investments in organizational capacity building, as well as human and social capital, may appear relatively small but, in the long-run, greater benefits and sustainability can be expected.

It is important to recognize the need for public and private-sector partnerships. The involvement of governments, the private sector, and donors contributes to the effectiveness of project implementation and reinforces project delivery. The involvement of the public sector ensures continuity, and the scaling up of successful projects. It also provides the private sector and donors with greater assurances of continuity, which in turn strengthens commitment.

Conclusions

As noted at the end of Chapter 1, this book addresses two sets of issues: based on literature and field experiences it promised to indicate the strengths and weaknesses of interventions and initiatives formulated to improve smallholder market access; and evaluate improving market access as an approach which contributes to bigger development goals. The purpose of the empirical Chapters 6–9 in Part 2 was to present evidence and draw lessons from the determinants of successes and failures of selected CFC/FAO projects in assisting smallholder farmers to participate (and/or participate on more favourable terms) in agricultural markets/value chains.

The results summarized here are consistent with the evidence from the literature on the importance of stimulating the market-participation capacity of smallholders. Without an enabling environment that enhances their natural, human, social, physical, and financial assets, smallholders do not have the appropriate incentives or opportunities to participate in markets. The results of the field surveys conducted in the four countries showed that project activities targeting the five market-participation capacity indicators described in the methodology section, did contribute to strengthening market access and linkages.

On the basis of the four country case studies, and using both qualitative and quantitative methods, a series of lessons and best practices in designing and implementing commodity development projects have been identified. First, survey data revealed that projects mostly targeted better-off smallholders, those with relatively better access to productive assets and suitable agroecological conditions. Only a few poor and very poor smallholders were selected as participants in the projects. This bias towards the better-endowed smallholders generates the best economic returns to the intervention, but it implies that the most marginal and poorest rural people were likely to be excluded as beneficiaries – although some benefits may have accrued to them through second-order effects, mostly in terms of increased opportunities in local labour markets. Smallholders are a heterogeneous group facing different types of constraints to market access and, as such, the initial objective of a commodity-development project should be to articulate clearly the nature of those constraints relevant to each category of smallholders, so that project execution is well-focused and likely to benefit the intended beneficiaries.

The analysis of the field survey data suggested that project activities addressing all five market participation capacity indicators contributed to strengthening market linkages. Further, improvements in smallholder market participation were associated with project activities that focused on extension, training and demonstrations, and support in building up productive agricultural assets. Gains in market participation were also correlated with the initial condition related to household and farm characteristics such as wealth, land size, asset ownership, and a favourable agroecological environment. Access to credit was found to significantly influence access to markets, highlighting the positive role of credit-support activities which constituted, in several instances, a core component of CFC/FAO projects. Favourable institutional and market environments were also factors predisposing towards project success.

The study surveys showed that technology adoption had a significant impact on smallholders. Yields per hectare were improved significantly in the case of projects involving crop activities (Ethiopia, Tanzania, and Peru), while milk yield per cow was increased twofold on average in Zambia. Project-specific activities were a factor behind the increases in productivity. These involved providing training to smallholders on good crop and livestock production practices, and provision of planting materials, market information, and subsidized fertilizers. In the case of Ethiopia, the project instituted a revolving loan fund to be used by the smallholders' cooperative for the purchase of inputs. As a result, smallholders managed to successfully grow and export green beans, at least for the lifetime of the project. In the case of the jatropha project in Peru, farmers learned new production techniques, including pruning, use of organic manure, bioinsecticide use, and utilization of bees to pollinate crops. Increases in earnings from commercial farming triggered additional incentives to generate production surpluses, notably during the initial years following the implementation of project activities.

Given the existing bias in the selection of participating smallholders, project activities and policy recommendations need to be specific to the targeted group. For better-off smallholders, priority should be given to areas addressing standards, quality, and export markets. In other words, policy interventions need to put the emphasis on marketing issues, which, from the survey data, appeared to be one of the main barriers to expanding market access. Note that government policies on prices and trade are most likely to have the largest impact on this sub-group of smallholders, given that most of them already participate, to varying degrees, in local, national, and to a certain extent, international markets.

For the poorest smallholders, priority should be given to activities that build up private productive assets and ensure the preservation of natural capital, so that poorly endowed smallholders are able to accumulate the assets necessary to sustain a commercially oriented farming strategy. Where access, and titling, to fundamental assets such as land and water can be improved, the potential increases in scale can provide a better platform for the less well-endowed smallholders. This can be the foundation for development through transfer of technology, skills in farm management practices (including sustainable land management practices), and access to credit and financial services. For the latter, the creation of a credit-revolving fund with farmers' organizations can assist smallholder farmers to integrate with input markets on a sustainable basis, as illustrated by the case study for Ethiopia. The four case studies showed there were considerable gains in social-asset building and networking as smallholders organized themselves into cooperative groups, associations, and unions.

The provision of reliable, affordable, and easy-to-access market and trade information is essential to sustain integration in both input and output markets. However, often more emphasis is attached to agronomic and production aspects. There is a pressing need to integrate marketing analyses into project activities at the initial stages of implementation so that smallholders, both women and men, are able to gain an understanding of business, and access and use market information including market analysis of current crop situation and outlook. A related issue is the ability of smallholders to assess and manage risks in the face of fragile ecosystems and volatile markets. Excessive risk discourages farmers from adopting new technologies and practices. The case studies evidenced a lack of consideration for risk management issues, including crop and market-outlet diversification, crop insurance schemes, financial inclusion, and development of on-farm genetic diversity. Systemic risk must be reduced through public investments into infrastructure, improving governance of agricultural markets as well as natural resources. Hence the importance of involving the public sector in the development of risk management outputs for commodity-development projects.

A key factor ensuring the success of a project is whether gains can be sustained beyond the lifetime of the project. There are two main lessons concerning sustainability that can be drawn from the case studies. First, gains are most likely to be sustained when project beneficiaries rely on more than a single commodity and/or a single market outlet. In the case of Ethiopia, there were major design and implementation failures when the participating exporting company halted its green-bean exporting operation, leaving smallholders with no export channel at the termination of the project. Similarly, in Peru, reliance on a single buyer rendered project beneficiaries, especially those without complementary activities, vulnerable to market shocks and power imbalances.

Second, special attention should be devoted to situations where a participating private firm holds monopoly or monopsony power since it may leave smallholders in a vulnerable or unsustainable position, as in the cases of Ethiopia and Peru. It is suggested that project appraisals should contain fully elaborated details on commitments made by value chain partners to enable each party to fulfil their obligations under the project, and beyond, with proposed contingency options in case of shocks. Sustainability of market access depends not only on the ability of smallholders to access input and output markets, but also on how these markets operate in the long term.

References

Amrouk, El M., Poole, N.D., Mudungwe, N. and Muzvondiwa, E. (2013). *The Impact of Commodity Development Projects on Smallholders' Market Access in Developing Countries: Case Studies of FAO/CFC Projects*. Rome, United Nations Food and Agriculture Organization. Retrieved 28 March 2017, from http://www.fao.org/docrep/017/aq290e/aq290e.pdf.

World Bank (2017). *Enabling the Business of Agriculture 2017*. Washington, DC, World Bank.

CHAPTER 11
Postscript: 'Going local' with development policies

There is a huge potential for smallholder agriculture to make a major contribution to overcoming the sustainable development challenges of the 21st century. Feeding a growing population through increasing smallholder participation in markets is one such contribution. For reasons of efficiency and best returns on investment in development interventions, policymakers and practitioners must learn from successful examples in order to 'upscale' the impacts of best development practice. Projects and programmes designed by public and other development organizations are not sufficient to meet the global challenges. It is essential to shape the institutional environment so that the autonomous initiatives of entrepreneurial households and firms can overcome entry barriers and respond to market incentives. The responsibility for shaping the institutional environment and policies tends to be located in the political centres of developing countries. Nevertheless, the centres often lack the detailed understanding of local people, contexts, and territories required to design and implement development programmes and projects. Alongside good communications between the respective levels, 'downscaling' decision making and responsibilities to provincial, regional, and district levels is necessary to address the challenges of the 21st century, and has important implications for governance, knowledge management, and capacity building.

Keywords: Sustainable Development Goals, governance, market participation, policy, knowledge, upscaling, downscaling, local, territorial, subsidiarity

Feeding the world

We began this book on market participation by arguing how significant the global smallholder agriculture sector is in terms of the number of people involved, and how farming is a way of life as well as a source of livelihoods for millions of poor people. In the opening chapter we stated the obvious, which is that the primary agricultural sector is the source of foodstuffs for the global population. We note here that the sector's contribution includes supplies of agricultural crops and livestock products to non-food markets such as fibres. It is food that grabs the attention. The demand for food is increasing in proportion with the growing population: about 7.5 billion people now, and possibly over 11 billion by 2100, with the greatest proportionate regional increase happening in Africa. The principal points of this book are to signal both the challenges and the potential for including smallholders in these

http://dx.doi.org/10.3362/9781780449401.011

agricultural markets, thereby meeting their development needs and, at the same time, making a much-needed contribution to sustainable food supplies and wider development objectives.

We highlighted the significance of heterogeneity – in brief, that not all smallholders may be willing or able to take advantages of opportunities for improved market access to become commercial farmers. Regarding inherent characteristics or attributes, many smallholders lack the necessary livelihood assets to commercialize, and, for some, there is a personal disinclination to farm commercially in favour of seeking other forms of livelihood and employment. There are the genuine barriers of remoteness, the small scale of enterprises, high business transaction costs and risks, inadequate services and technologies, and organizational failure, which are among the external and contextual factors which limit smallholder market participation. In Chapter 2 we emphasized the significance of individual and household behaviour in relation to markets, and particularly how farmers make decisions not only about production costs, but also the transaction-costs associated with agribusiness.

But we have also noted that many small-scale farmers are already linked into markets for inputs, products, and services – as suppliers and as consumers – on their own initiative, without outside intervention. Again in Chapter 2, we identified trends in policy approaches to stimulating market participation that have led to the prevailing value chain paradigm. From the literature and from the case studies in Part 2, we have seen that, where entrepreneurial potential is frustrated by surmountable barriers, the difficulties of which we have a good theoretical understanding, well-designed external interventions can facilitate viable linkages to agri-food value chains. Such projects tend to be costly, and the beneficiaries few, so there is always a desire to multiply impact through efforts to upscale and replicate, by applying the lessons learned to a much wider population and therefore getting a better return on the project investment. Such multiplication processes are important, but costly projects and external interventions cannot match the potential opportunities. Individual households and businesses in existing market systems must be encouraged and shaped within the context of improving local agroecological and human situations, commercial opportunities, and local institutional environments.

It is under these circumstances that appropriate support and services should be delivered to the wider population. Chapters 3 and 4 built on Chapters 1 and 2 by exploring in depth the fundamental importance of providing financial services and risk management for smallholders who wish to exploit market opportunities. We stressed the need for diagnosis of the business challenges and for the design of comprehensive services to enable farmers to overcome the market-access barriers, with a significant role for the public sector, and, optimally, in relationship with private-sector provision of finance, marketing and business skills, and human-capital formation. The cases in Part 2, Chapters 5–10, illustrated project-design and implementation issues and how these challenges were addressed – to a greater or lesser extent.

In this postscript, we draw together a number of threads to consider how market access fits into the big picture of development. We begin with overview comments on the macro, global scale of agricultural development, and end with more precise local recommendations. There may appear to be a contradiction with the conclusions of Chapter 10 which imply the potential of lessons for upscaling good practice, but this argument comes with a warning. It constitutes a plea – or pronouncement, or even a provocation – as a corrective to the conclusion that market-access experiences can necessarily be replicated and upscaled into policy. It is a framework for action, or a manifesto, to scale down policymaking from the macro to the micro, meaning that policymaking should be downscaled to the local level, building on local knowledge and understanding of specific people and contexts. Actually, it is not a question of either upscaling or downscaling, but of rebalancing the onus of responsibility for development policy, programmes, and projects towards the local context and away from the central axes of development decision making.

A macro perspective

The Sustainable Development Goals (SDGs) provide the current framework for understanding and formulating strategies. But strategies are only the basis for policies which then have to be operationalized through programmes, projects, and local processes. As we move from the outer ring of Figure 1.1, embracing the macro issues of development, towards the inner ring, which includes the livelihoods of poor agricultural smallholders who are our focus, in a geographical sense we downscale from the global to continental, to national, to regional, to district, to local 'micro' issues.

Food quantity

There are huge challenges in feeding the world. While innovative food sources such as insects, bacteria, and seaweed, among others, will be developed and may become acceptable to consumers over the 21st century, there is still a need to increase areas of production, increase productivity, and change food sources and diets. There are limited areas for the expansion of food production, but those there are lie largely in developing regions. The data reviewed in Chapter 1 illustrated the productivity gap between advanced-country agricultural systems and these developing regions, offering considerable opportunity for increasing food supplies.

Many developing regions are characterized by fragile ecosystems: poorer soils than the productive temperate regions, extreme temperature regimes, and suboptimal rainfall patterns. Meteorological extremes are already increasing; this is likely a consequence of climate change and, so far, international agreements have not been effective in halting the rate of global warming. While there is need for large-scale investments to exploit the constrained potential of these relatively unproductive lands, they are precisely where smallholder agriculture also is an important mode of production.

As populations move in response to changing lifestyle preferences and livelihood opportunities, poor populations will be found more often in urban areas than rural areas, increasing the demand for infrastructure and the market functions of storage, transport, and processing. Relatively high levels of male migration, and perhaps increasing whole household shifts, will mean fewer human resources for the rural economy. Targeting agricultural development programmes towards women will become even more important. There will be an increasing requirement for a shift from labour-intensive production to systems that make more use of capital – at the same time reducing the drudgery that often is a disincentive for young people to commit to agriculture.

Sustainability

The Sustainable Development Goals, concerning the many issues challenging all nations, have attracted the attention of the global community. Among other things, agriculture and downstream sectors make a large direct contribution to the problems as well as potential solutions, not least to global warming through direct and indirect greenhouse-gas emissions. For agriculture and rural activities in general, sustainability implies proper management of the natural resources base: water, soils, biodiversity, pastures, woodland and forests, and air. Increasing utilization of more sophisticated technologies must be consistent with – indeed, supportive of – sustainable intensification.

For development practitioners, difficult questions surround these challenges: how much will these changed practices happen autonomously, guided by the invisible hand of markets? How much will they require a policy response from governments and supranational organizations to address market failures? How much collective will to make an effective policy response is there in the international community? We will see by 2030 whether the SDGs have been the right ones and how successfully policies have been operationalized. It will become evident how much progress has been achieved in orienting agriculture towards the quality and quantity of food delivered to consumers that will be needed to meet the 2030 Goals.

A sectoral perspective

Health and food quality

Market participation is not an end in itself, but for most smallholders is a means towards the end of earning a living and overcoming poverty challenges. We noted above the importance of production which sustains the natural environment. Besides this stewardship function, agriculture can directly address major health challenges. Agriculture can make a direct contribution towards reducing the undernutrition that causes costly and distressing human, social, and economic underdevelopment. Agriculture's contribution as an input to the local and global food industries is also linked to the manufacture and distribution of cheap foodstuffs, and eventually to the poor quality of diets that cause the other half of the double burden of malnutrition – overnutrition.

The global food system is a major driver of the demand for specific crops and livestock products, the overconsumption of which is implicated in the rise in non-communicable diseases attributed to processed meats and dairy products, sugar, and palm oil among other foodstuffs. This suggests that smallholder agriculture should be not only efficient and profitable but also sensitive to consumers' nutritional requirements, particularly in respect of micronutrients. There are theoretical reasons, and plenty of evidence, to expect smallholders to be able, efficiently and sustainably, to supply value chains for high-nutrient quality foods such as fruits and vegetables. Smallholders can exploit potential competitive advantages in supplying local markets with foods of high nutrient quality, whose consumption in many developing regions needs to rise to meet nutritional quality objectives.

Economic multiplier effects

We have commented that the value chain concept is central to contemporary agri-food policy interventions by international and governmental organizations, and many development NGOs. The attraction is the whole-chain focus which can encompass the complex range of activities necessary to bring primary products to market, and the vision for creating economic opportunities before and beyond primary production. These include involving small- and medium-scale enterprises (SMEs) in the marketing of input supplies, plus employment in the transportation of primary products, processing, manufacturing, distribution, and retail. SMEs beget SMEs as incomes are recycled to supply the demand for a wider range of goods and services, and so the local economy grows.

Again, the theory and evidence we have presented in earlier chapters show that the more entrepreneurial members of the rural community, including smallholder farmers, can overcome the market-entry barriers and integrate upstream and downstream functions as new economic activities. New skills are learned, incomes are raised from exploiting potential economies of scope from a diversified livelihood portfolio, and risk can be identified, managed, and reduced. Conventional small-scale agriculture has a declining attraction for young people, and consequently farming – globally – is characterized by an ageing population of farmers. But young people can be attracted into sectors that are made of more technically challenging, capital-intensive enterprises that effectively balance risk and reward.

Infrastructure and more

The growth of commercialization among smallholder farmers, as in any other economic activity, depends on infrastructure and logistics: provision of water relies on physical capital investment; manufacturing and distribution rely on power and roads and bridges; market supply and demand for all sorts of goods and services depend on timely and accurate information using hard telecommunications and soft knowledge. The growing body of evidence that access to markets is a driver of better nutrition even in rural areas is an illustration of how health and well-being are linked to the economics of the market and not just household food production. Investment in education and training or, in a broader sense, human-capacity building, underpins much of human progress. Twenty-first century information technologies can make knowledge and access simpler, delivering them more equitably and more efficiently than at any other time in history. Again, to argue this is not novel, but more a rediscovery of the complex and comprehensive characteristics of the interventions necessary to tackle development problems.

Policy interventions and tensions of scale: going micro

We have noted that the people and households at the centre of Figure 1.1 are heterogeneous in terms of their assets and inclinations. The natural contexts are also heterogeneous and often cross borders which do not respect agroecological resources such as soils and water basins, nor environmental, social and political entities such as markets, ethnicity, conflict, migration, zoonoses, antibiotic resistance, and pandemics. The point is that development challenges, including improving market access and participation, are not confined within political limits: the scale is trans-territorial, and not a function of the nation-state. There are other trans-territorial phenomena that can contribute to development of trade and food systems, transfers of knowledge and information. Distinguishing the implications of trans-territorial transfers and barriers needs skill and understanding of context. The implication is that governance of the SDGs should not be taken to be equivalent to the governance of the nation-state by national policymakers, but needs to be framed within a context of 'trans-national territoriality'.

Second, there is a tension between the scales of thinking which frame the macro strategies and the design of policies, programmes, and projects – in our case, of market participation – at a micro level. The features of diversity and locality must impel development policymakers and practitioners to be responsive to context, as highlighted in the literature and in the comments on the case studies in Part 2. This much is often noted in thinking, but less often respected in the doing. Even the cases reviewed earlier demonstrate that successful market participation can be achieved where there is understanding of the human, social, political, and economic context. The failures in project design and implementation can be attributed, at least in part, to lack of that same understanding.

Again, this is not a new finding, but is perennial and persistent. Do central policymakers have too much influence over designs for local contexts, and do remote donors who operate at the macro level prescribe too much the interventions that take place at the local level? After decades of post-colonial development and aspirations towards participatory development design and management, are the projects still being designed and authorized at the wrong level?

Exploring these ideas in a specific context may help emphasize the argument. Recent policy-oriented work in agriculture and nutrition in Afghanistan, to which we referred in Chapter 1, identified significant gaps between stakeholders concerning actual policies, programme design, and project implementation (Poole et al., 2016). Most importantly, doubts were expressed by provincial stakeholders about the level and validity of knowledge of the people at the centre. Development practitioners and some public-sector officials considered that people from the centre – that is, the decision makers largely based in the capital, Kabul, and many foreign experts – were ill-informed about the provincial realities. Interviewees suggested that policies were misguided, and that knowledge exchange between the different levels of decision making and implementation was deficient.

Not all countries are as heterogeneous as Afghanistan, which is an extreme challenge for development work. Afghanistan has a very wide diversity of ecological regions and of ethnicities, traditions, and patterns of livelihoods. There is extreme remoteness and there are major physical barriers to efficient logistics and the exchange of knowledge and information. The delivery of public services for health, education, and training in this complex natural and social environment is arduous. Natural resources (besides the mineral wealth) are limited. Parts of the country are prone to natural disasters such as avalanches, landslides, and earthquakes. The climate is already changing and the incidence of adverse effects is likely to be severe. Above all, the ongoing conflict exacerbates the impacts of all the difficulties, and contributes directly to the provincial–central gap.

The specific weaknesses identified in the Afghanistan policy environment are not unique or new (Poole et al., 2016: 86–8):

- little integration of development policies across sectors such as emergency relief and longer-term development, agriculture and nutrition, education, markets, infrastructure and knowledge management, and gender;
- policies reported to be often ill-designed through top-down processes, in part driven by ill-informed international organizations and experts;
- imperfect coordination activities among myriad policy-formulating bodies, funding and implementation partners caused many inconsistencies;
- lack of capacity and resources within government ministries and departments along with poor infrastructure.

In addition, Afghanistan is racked by huge security concerns which remain major barriers to progress. These may not be typical of developing regions, but the proportion of poor countries that suffer conflict is not coincidental. The lessons are likely to have a wide application in fragile regions.

Detailed knowledge of smallholders and markets is a local phenomenon. In the case of Afghanistan, and other developing economies which have appeared throughout earlier chapters, a subnational governance architecture can offer a structure that could allow the decentralization of development responsibilities to the regions and provinces, to local policymakers and practitioners who know the localities better than national policymakers and international donors and experts based in the capital. Decentralizing responsibilities does not only require a governance structure but is also conditional upon having the appropriate human capacity at the regional, provincial, and district levels. While insecurity persists, staffing remote areas is problematic. In Afghanistan, peace-building is an absolute necessity but, given greater security and stability, downscaling development responsibilities to the local context is an approach that could respond to the plea made in this postscript. Coordination between provinces also offers an approach to the territorial nature of the development context.

Transnational territorial approaches to apparently intractable local development challenges may seem idealistic and improbable, but other approaches have failed to make an impact. However, there is a governance principle of subsidiarity to which we can appeal for support: adopting the principle of subsidiarity enables decision making and intervention to be undertaken at a level as close as possible to the citizen or stakeholder, i.e. as little centralized as can be made to work competently.

The approach proposed here suggests the following three methods to downscale actions for market development and other broader responsibilities, which may find echoes in other contexts:

1. *Governance*. Greater decentralization of policymaking to provincial levels. Devolution requires that the scale of operations strikes a balance between the availability of technical expertise and local capacity. Provincial/regional-level governance should result in policies that are likely to be context-specific and appropriate.
2. *Knowledge management*. Improved information flows of local knowledge between central, provincial, and local government structures. Increased use of local governance – both line ministries and the governors' offices – and effective use of IT are mechanisms to bridge the current poor lines of communication and information flow between central and provincial/regional levels.
3. *Investment in local government*. It is essential to build capacity and overcome the 'brain-drain' from the regions, as individuals with capacity tend to be recruited to work under more attractive employment conditions with central government, (I)NGOs, and UN agencies. Human capacity must be built at decentralized levels to increase local competence.

The final question is where to start downscaling? By all means, let national governments re-envision policy processes by adopting decentralized and territorial approaches to development. But a significant first move is needed by policy and decision makers, and researchers within the architecture of international development organizations, to engage directly with the individuals and households who are at the centre of sustainable development, and thereby gain a comprehensive and contextualized understanding of local realities at first hand.

Reference

Poole, N., Echavez, C. and Rowland, D. (2016). *Stakeholder Perceptions of Agriculture and Nutrition Policies and Practice: Evidence from Afghanistan. LANSA Working Paper Number 9*. Chennai, India, Leveraging Agriculture for Nutrition in South Asia (LANSA), MS Swaminathan Research Foundation. Retrieved 24 May 2017, from http://ims.ids.ac.uk/sites/ims.ids.ac.uk/files/documents/Mapping%20stakeholder%20perceptions%20%20Afghanistan%20on%20template_0.pdf.

Index

Page numbers in *italics* refer to boxes and tables.

Afghanistan: policy stakeholders research 9–10, 189–90
Africa Progress Panel *14*
agribusiness 54
agricultural development
  challenges in fostering entrepreneurship 12–15
  commercialization 16–25
  and poverty reduction, contributions and limitations 8–12, 25–7
  terminology 27–9
All African Caribbean Pacific Agricultural Commodities Programme (AAACP) 72, 76, 83
Amrouk, E.M. et al. 17, 20, 118
Angelucci, F.
  and Conforti, P. 72, 73
  et al. 14
Arias, P. et al. 5, 6, 16–17, *21*, 22
assets *see* livelihoods/asset-based approach

Bangladesh: rice trade 65–6
bargaining models 43
Barrett, C.B. 21–2, 36, 40, 42, 44, 45, 47–8
behaviour
  embeddedness and business systems 59
  household 42–3, 184
  information challenges in financial service provision 68–9
  marketing 47, 90, 91, *95*
  Rational Economic Man 46
  risk management 96
biofuel production *see* Peru: oilseed production for biofuel
Borzoni, M. and Poole, N.D. 106
business systems 59

capacity development 178
centralization and decentralization 190–1
cheating: cartel-like marketing agreements 90, *91*
children 42, 43, 52
ChimpReports 24–5
climate change 14, 72, 96, 98, 185
Coase, R.H. 46, 47, 48, 53–4
Coles, C.
  et al. 58
  and Mitchell, J. 57

collective and individual decision making models 43
collective organization *41*, 74, 109
commercialization 16–25
commodity development projects
  cases 118–21 (*see also specific case studies by country*)
  conclusions 178–81
  lessons learned 126–7, 166–78
  recapitulating theory of change: expected relationships 165–6
  recommendations for selecting project-participating smallholders 176–7
  research approach and framework 121–4
  research methodology and study limitations 124–6
  shortcomings in project design and implementation 172–5
Common Fund for Commodities (CFC)/FAO-project case studies 118–21
Commons, J.R. 45–6
competitive strategies 54
contracts
  hedging price variation 91–2
  role of 93–5
convertibility of assets 52
coordination
  Afghanistan 189, 190
  horizontal and vertical 58
  market economies 37
  and risk management 90–1, 92, 93, *95*
  state and private enterprises 139
  value chains 53–4, 108
coping strategies in risk management 88, 93
credit
  access to 165–6, 168, 169, 171–2, 179, 180
  lack of access to (case studies) 153, 154, 159–60
  and risk management 88, 89, 93, 101, *105*
  see also financial services
Crow, B. 65–6
cultural capital 50
  erosion of 52

dairy *see* Zambia: strengthening productivity and competitiveness of smallholder dairying
decentralization 190–1

decision making
    production, storage, and marketing 89–91
    see also households
Department for International Development (DFID), UK 38
development goals see Millennium Development Goals (MDGs); Sustainable Development Goals (SDGs)
development policies
    feeding the world 183–5
    macro perspective 185–6
    market systems to value chains 35–42
    micro perspective 188–91
    sectoral perspective 186–8
distress sales 89
donors 35, 60, 67, 100
    ideologies and policies 37
Donovan, J.
    et al. 146
    and Poole, N.D. 45, 51, 75, 107
    and Stoian, D. 49, 51
downscaling 183, 185, 190–1

economic decision making 42, 43–4
economic multiplier effects 187
efficiency *17*, 25, 183, 187, 188
    case studies 136, 162
    collective organization *41*
    cooperatives 59
    failure of market liberalization 36
    financial services 75–6
    process upgrading 20
    risk management 85, 93, 111
    transaction cost economics (TCE) 47
    unitary model 42
    value chains 54, 56
employment potential of agriculture *14*
Enterprise Finance Guarantee (EFG), UK 75
entitlement relations 48–9
entrepreneurship, challenges in fostering 12–15
equity, leveraging 75
erosion of assets 52
Ethiopia: diversification programme for vegetable export development 119–20, 129–37, *167*, 170, 174–5, 178, 179–80, 181
    access to land 131
    changes in agricultural assets 131–2
    conclusions 136–7
    costs of quality 134
    downstream activities 133–4
    financial assets 133
    household asset building 131
    project design and implementation 129–30, 135–6
    project services 132, *133*
    questionnaire sample 130
    sustainability 134
    targeting 135

Ethiopia: teff cultivation and marketing *19*
exports 16, 30, 45, 56, 92
    Ethiopia: teff cultivation and marketing *19*
    Madagascar: French beans *109*
    restrictions 91
    Sudan 119–20
    see also Ethiopia: diversification programme for vegetable export development; entries beginning international
external environment 12, 25, 44, 48, 53, 55, 59
    Tanzania (case study) 149, 155–6

Fafchamps, M. and Minten, B. 93–4
Famine Early Warning System Network (FEW NET) 96
financial assets 50, 123, *167*
financial services
    challenges for rural enterprises 71–2
    complementary and supporting services 75–7
    information challenges 68–71
    innovations for farming 73–7
    and level of market sales 22
    newer services 67, *68*
    risk in agriculture and rural economy 72–3
    role of 65–7
    supply 73–5
    and timing of marketing decisions 89
    and transaction costs 68–73
    see also credit
flowers for international markets 108–9
Food and Agriculture Organization (FAO) 5, 14, 29
    and CFC-project case studies 118–21
food production
    and population growth 4, 5–6, 183–4
    and productivity 6, *7*
food quality 186–7
food quantity 185–6
food security 4, 5, 9, 12–13, 14, 16, 18, 24–5, 27, 30
    financial services 66
    risk management 104, *105*
    role of governments 111
    Zambia 157
'food sovereignty' approach 18
forward contracts/forward sales 92, 93
frameworks
    development goals 8–9
    ethical 23
    institutional *41*
    livelihood assets *51*
    research 121–4
    role of governments 97, 111
    strategic 38
    value chains 53–6

INDEX

fresh fruits and vegetables
   for domestic consumption 105–6
   for international markets 108–9

García-Martínez, M. and Poole, N.D. 107
gender *see* households; women
Ghana 47, 74
   formal contracts 95
Global Index Insurance Facility (GIIF) 99–100, *101*
governance 190
government roles and interventions
   finance 66–7
   insurance 87, 97, 99, 100–1
   price stabilization 91
   public-private sector partnerships 74–5, 76–7, 87
   risk management 95–7, 111
   staple foods for domestic consumption 104–5
green beans
   from Madagascar *109*
   see also Ethiopia: diversification programme for vegetable export development
group lending 67
Guyer, J.L. 50, 51, 52, 53

Haddad, L. et al. 42, 43
Hardaker, J.B. et al. 72
Harper, M. et al. 37, 118
health
   and food quality 186–7
   nutrition-related indicators 7, *8*
Helmsing, A.H.J. and Vellema, S. 24, 44, 58, 59–60
heterogeneity 13, 179, 184, 188
   Afghanistan 189
   households 20–3
households
   economic decisions 42, 43–4
   heterogeneity 20–3
   and livelihoods 48–53
   management decisions 43–4
   market decisions 44–5
   modelling 42–3
human assets 49, 123, *167*
Human Development Report 24

IFAD 15, 50
impacts *see* commodity development projects
index-based insurance *101*
   and indemnity, differences between 98
India 23
indicators 28
   development goals 8–9
   health and nutrition-related 7, *8*
   impact of projects 165–9, 178–9
   livelihood asset 122–3

   market-related 123–4
   of natural capital 52
   smallholder participation in marketing 18
   World Development 5–6
individual and collective decision making models 43
inequality and opportunity, market exclusion and inclusion 23–5
information
   finance services 68–71
   knowledge management 190
   risk management 96, 97
infrastructure 187–8
innovation 15, 20
   financial services/insurance 67, 73–7, 97, 100–1
   institutional 40–1
   livestock feeding technologies 121, 158, 170–1
   as market-related indicator 124
   risk management 110
institutions
   innovation 40–1
   role and transaction costs 45–7
insurance
   for agricultural development 97–101, 97–8
   compulsory crop 74
   government provision 100–1
   government regulation and legislation 97, 99
   risk layering 84–5, 86–7
   types 98–100
interactions and trade-offs 51–2
international commodity agreements 67, 90
international fresh produce markets 108–9
international fresh produce-chains 102–4
International Research Centre, Canada 57
internationally traded commodities 106–7
internationally traded speciality commodities 107–8
interventions
   case studies 129–30, 146, 155–6, 157, 162
   impact of projects 117–18, 122, 126–7, 169, 176–7, 178–80
   see also development policies; government roles and interventions
investment 51
   in local government 190

Jaffee, S. 129
   et al. 81–2
   Siegel, P. and 82, 83
Jatropha *see* Peru: oilseed production for biofuel

Karamojong people, Uganda 24–5
Karembu, M. 15
knowledge management 190

Latin America 23
livelihoods/asset-based approach 48–53, 122–3
 linked financial services 75
 project impacts *167*, 170–2 (*see also specific case studies by country*)
Lloyd's Micro Insurance Centre 99, *100*
loan guarantees 74–5
local context 25
local development *see* development policies
local financial resources 37
local food systems 18
local government, investment in 190
local and international/global integration 16, 24, 48, 60, 89
local markets 18, 23, 40, 48, 89
local and particular analysis and interventions 21, 40
local producers 17, 92
local suppliers 16, 74

Madagascar: French beans from *109*
making markets work for the poor (MMW4P/M4P) 37–40
Mali: social capital development 74
management
 decisions 43–4
 knowledge 190
 see also risk management
Marion, B.W. and NC117 Committee 46
market access
 and constraints 6, 20, *21*
 MMW4P/M4P 37–40
market exclusion and inclusion 23–5
market liberalization 36
market monitoring 18–19
market participation 16–20
market systems policies 35–42
market-related indicators 123–4
marketing
 adequacy of (case studies) *133*, 153
 behaviour 47, 90, 91, *95*
 decision making 89–91
 indicators of smallholder participation 18
microfinance 67, 75, 76–7, 93, 99
microinsurance 98–9
 and traditional insurance, differences between *100*
migration 13
Millennium Development Goals (MDGs) 8, 40
Miller, C. and Jones, L. 73, 75–6
Minten, B.
 et al. 19, 108, *109*
 Fafchamps, M. and 93–4
multi-sectoral problems 9–12

natural assets 49, 123, *167*
natural shocks 52, 72, 106
New Home Economics 42–3

New Institutional Economics (NIE) 46–7, 54
NGOs 56, 60, 67
Nicaragua: linked financial services 75
Nigeria: private-sector contractual finance 74
North, D.C. 48, 71, 92
nutrition and health 6–7, *8*

oilseeds *see* Peru: oilseed production for biofuel
Old Institutional Economics (OIE) 46
Overseas Development Institute, UK 57

partnerships
 chain-based 59–60
 public-private sector 74–5, 76–7, 87
people perspective 14–15
Peru: oilseed production for biofuel 120, 139–47, *167*, 170
 conclusions 145–6
 demographic details 140
 impacts: strengths and weaknesses 144–5
 knowledge management and technology 142
 productive capacity 141–2
 programme design and implementation 139–40
 project services 141
 public support 143
 scale of production 141
 sustainability 43
physical assets 50, 123, *167*
policy approaches and interventions *see* development policies
Poole, N.D. 40, 53, 76–7, *95*
 Borzoni, M. and 106
 and de Frece, A. 40, *41*
 Donovan, J. and 45, 51, 75, 107
 et al. 9–10, *12*, 15, 18, 21, *22*, 23, 25, 36, 50, 76, *95*, 108, 189
population growth
 global 4, 183–4
 rural 6
Porter, M.E. 54, 56
post-colonial period 35, 65–6
poverty reduction
 and agricultural development, contributions and limitations 8–12, 25–7
 livelihoods approach 48–53
 making markets work for the poor (MMW4P/M4P) 37–40
price variability *see* risk management
private-sector contractual finance 74
processing
 financial services 74, 76
 impact of projects 121, 123, 125, 170
 risk management 94, 102, 104, 107
 see also Peru; Tanzania; Zambia
production, storage, and marketing decisions 89–91

production costs 46, 47, 48, 82, 143, 150, 184
productivity 16–17
　food 6, *7*
　see also Zambia: strengthening productivity and competitiveness of smallholder dairying
project/programme design
　case studies 129–30, 135–6, 139–40, 149–50, 157
　shortcomings in 172–5
Provost, C. and Jobson, E. *19*
public-private objectives 104, *105*
public-private sector partnerships 74–5, 76–7, 87
public-sector *see* government roles and interventions

qualitative impact evaluation 121–2

remittances 13, 75
remoteness 22, 86, 102, 105, 126, 155, 184, 189, 190
research
　Afghanistan: policy stakeholders 9–10, 189–90
　approach and framework 121–4
　methodology and study limitations 124–6
　value chains 55, 56–7
rice trade, Bangladesh 65–6
risk assessment 83
　identifying and mapping risk 83–4
　risk layering 84–7
risk management 87, 110–11
　coping strategies 88, 93
　economic risk and types of value chains 101–9
　effects of production, storage, and marketing decisions 89–91
　and finance 73
　hedging price variation 91–2
　role of contracts 93–5
　role of governments 95–7, 111
　see also insurance
Rogers, T.S. et al. 16
Rural Development Report 2016: Fostering Inclusive Rural Transformation (IFAD) 13, 50
rural population
　and food production 5–6
　migration 13

Sahel region, West Africa
　households and management decisions 43–4
　livelihoods and ecology *12*
　self-insurance 85
Sen, A. 23, 48–9
side-selling: cartel-like marketing agreements 90, *91*

Siegel, P. and Jaffee, S. 82, 83
sisal *see* Tanzania: sisal product and market development
Small Island Developing States (SIDS) 7, 73
smallholder agriculture, significance of 3–8
Smith, A. 94
social assets 50, 123, *167*
social capital 50
　development 74
Spain: formal contracts *95*
The Springfield Centre 38, 39
stakeholder commitment 177
staple foods
　for domestic consumption 104–5
　markets 44–5
State of Food and Agriculture 2016: Climate Change, Agriculture and Food Security (FAO) 14
subsidiarity principle 190
supply-chain risk management (SCRM) 82
sustainability 52–3, 186
Sustainable Development Goals (SDGs) 3, 23, 40–1, 185, 186
　and interrelationships 8–12, 26
Swenson, L. 81

tail risk 86–7, 96
Tanzania: sisal product and market development 120, 149–56, *167*, 170, 175
　access to land 152
　adequacy of marketing 153
　agricultural assets 151–2
　conclusions 154–6
　financial assets 153–4
　household assets 151
　programme design and implementation 149–50
　project services 152–3
　questionnaire sample 150–1
　social assets 154
targeting 13, 53
　design of project interventions 126, 172–3, 176–7, 179–80
　middle-income farmers 135
　women 186
technical services 73, 139, 149
terminology 27–9
territory/territoriality 59, 60
　trans-territoriality 188, 190
theory of change 12–13, 28, 29, *122*, 165–6
threshold
　concept 176–7
　minimum 28
　risk/risk management 47, 68, 109
trans-territoriality 188, 190
transaction costs 36–7, 40
　farmers' 47–8
　and finance 68–73
　role of institutions and 45–7

Uganda: Karamojong people 24–5
UN
  Human Development Report 24
  World Food Programme (WFP) 73–4
  see also Food and Agriculture
    Organization (FAO)
upgrading 58
upscaling 24, 40, 58–9, 60, 178, 183, 185
USAID: Famine Early Warning System
  Network (FEW NET) 96

value addition 54, 55, 102, 144
  Zambia (case study) 157, 160–1, 162
value chains 20, 24, 40–1
  approaches and definition 54–5
  concept development 53–6
  contextual factors *11*
  economic multiplier effects 187
  financing 73–4, 75–6
  interventions 56
  participation: principles and determinants
    58–60
  research 55, 56–7
  supply-chain risk management (SCRM) 82
  upgrading 58
  see also risk assessment; risk management
Via Campesina movement 18
vulnerability and risk management 82

warehouse receipt systems 104, *105*
weather
  climate change 14, 72, 96, 98, 185
  effects 106
  insurance 98

Wehling, P. and Garthwaite, B. *105*
Weightman, B. 4–5
Wiggins, S. and Keats, S. 17, 18, 26
Williamson, O.E. 46, 47, 54
women
  participation in decision making 43
  smallholders 4, *5*
  targeting 186
World Bank 4, 16, 20, 38–9, 87, 104, 111
  World Development Indicators 5–6
  World Development Report 2007 (WDR)
    21, *22*
World Food Programme (WFP) 73–4
world-systems theory 54

young people 14–15, 169, 186, 187

Zambia: strengthening productivity
    and competitiveness of smallholder
    dairying 120–1, 157–63, *167*,
    170–1
  changes in number of dairy cows 161
  conclusions 162–3
  household and agricultural
    assets 159
  improvement in market participation
    160–1
  productivity measures 161
  project context 157–9
  project design and implementation 157
  project services 159–60
  viability of milk collection
    centres 161
Zanello, G. et al. 47